Praise for *Redfield's Warning*

"Former CDC Director Robert Redfield, MD provides us with a riveting account of his experience in the eye of the COVID-19 storm, detailing the internal deliberations, policy debates, and successes and failures of federal officials. As with COVID-19, the next deadly pandemic could come from a lab, where scientists engage in dangerous 'gain of function' research: the genetic manipulation of pathogens to enhance their transmissibility and virulence, making them deadly to humans. Redfield insists that Washington must make biosecurity a top priority. Our safety depends on it."

—**Robert Emmet Moffit, PhD, senior research fellow at The Heritage Foundation and former principal deputy assistant secretary at the US Department of Health and Human Services**

"Dr. Robert Redfield is a visionary. I was working on the Bellevue wards treating AIDS patients in the mid-to-late 1980s when he engendered controversy by stating that HIV could be transmitted heterosexually. When it came to Covid, he was one of the first and loudest voices to state that Covid likely came from the lab, and since then, many others have fallen in line behind his leadership on that likelihood. In this book, *Redfield's Warning*, he again shows his great vision when he warns of the next pandemic, which he knows is coming."

—**Marc Siegel, MD, clinical professor of medicine and medical director of Doctor Radio at NYU Langone Health; Fox News senior medical analyst; and author of the upcoming book *The Miracles Among Us***

"Dr. Robert Redfield has written a book that is brilliant, bold, reflective, and full of recommendations for the future. His life story is one of great consequence, vision, and commitment to the mission of disease detection, and prevention, and to the pursuit of cures. Redfield faced opposition at many turns in his storied career beginning with his pioneering research into HIV and AIDs. The lessons of AIDs were reflected in his analysis and approach as CDC director of the massive threat of COVID-19, which he grasped earlier than virtually any other public health expert. Bob Redfield is an American visionary, patriot, and hero."

—David L. Asher, PhD, senior fellow at Hudson Institute

"Dr. Redfield is a truthteller, a former insider who knows it all. He lays out a stark reality, and I am using his words: 'The days of believing authorities because they are in an authoritative position are over.'"

—Gavin de Becker, bestselling author, *The Gift of Fear*

REDFIELD'S WARNING

WHAT I LEARNED (BUT COULDN'T TELL YOU) MIGHT SAVE YOUR LIFE

DR. ROBERT R. REDFIELD

Former CDC Director

Skyhorse Publishing

Copyright © 2025 by Robert R. Redfield

All Rights Reserved. No part of this book may be reproduced in any manner without the express written consent of the publisher, except in the case of brief excerpts in critical reviews or articles. All inquiries should be addressed to Skyhorse Publishing, 307 West 36th Street, 11th Floor, New York, NY 10018.

Skyhorse Publishing books may be purchased in bulk at special discounts for sales promotion, corporate gifts, fund-raising, or educational purposes. Special editions can also be created to specifications. For details, contact the Special Sales Department, Skyhorse Publishing, 307 West 36th Street, 11th Floor, New York, NY 10018 or info@skyhorsepublishing.com.

Skyhorse® and Skyhorse Publishing® are registered trademarks of Skyhorse Publishing, Inc.®, a Delaware corporation.

Visit our website at www.skyhorsepublishing.com.
Please follow our publisher Tony Lyons on Instagram@tonylyonsisuncertain.

10 9 8 7 6 5 4 3 2 1

Library of Congress Cataloging-in-Publication Data is available on file.

Hardcover ISBN: 978-1-5107-8505-2
eBook ISBN: 978-1-5107-8506-9

Cover design by Brian Peterson

Printed in the United States of America

In memory of all the men, women, and children who died of HIV and COVID-19, and for all those suffering from long COVID and from post-COVID mRNA vaccine injury. May science refocus its efforts to discover effective therapy for both.

ACKNOWLEDGMENTS

I want to thank all the patients I had the privilege to serve and for all they did to teach me the art of medicine. I also want to thank all my colleagues who I worked with over the years at Walter Reed, University of Maryland, and CDC. They too are responsible for teaching me science, medicine, and public health. Most importantly, I want to express my deepest love and admiration to my life partner, Joy, for the past fifty years. She always stood steadfastly by my side and enabled me to serve. Joy and I have been blessed with six children. Unfortunately, John Paul, our first son died, but God granted us five more children—Robert, Jennifer, William, Daniel, and Patricia—each of whom Joy and I are very proud of the people they have become. We have also been gifted with fourteen grandchildren: Jack, Max, Ella, Johnpaul, Riley, Sophie, Joey, Anna, Tessa, Duke, Reed, Ford, Charlie, and Alice Joy. We thank them for the joy they bring to our life each day. Our prayer for each of them is that they choose to embrace the wisdom of Mother Teresa: "The fruit of silence is prayer. The fruit of prayer is faith. The fruit of faith is love. The fruit of love is service. The fruit of service is peace." May each of them pursue a life of service and be rewarded with a life grounded in peace.

I would like to thank those who made Redfield's Warning possible. My deep appreciation to the team at Skyhorse Publishing, especially Tony Lyons and Hector Carosso. Also, a special thanks to Emily Fairbairn for her support. Finally, I want thank David Fisher, a New York Times bestselling author, for his skill in converting my life stories and experiences into an important and readable book.

ONE

Let's begin at the end: How many Americans are going to die in the coming pandemic? Three million? Four million? Pick a number. It's probably low.

At the beginning of Covid, as director of the US Centers for Disease Control and Prevention, I got an estimate from expert modelers that as many as 2.2 million Americans were going to die. I remember sitting at my desk looking at that number: 2.2 million dead Americans. My wife Joy and I believed there was a significant chance we would be among them.

The country was very fortunate. The virus only killed 1.2 million people. Imagine that, considering the country fortunate because only a million people died. But that's only a death rehearsal for what is coming.

No one can predict the source or the timing or the multimillion-person death toll of the next pandemic, but it is coming.

The virus that causes it might literally be thousands of years old. Within the last few years, researchers have revived ancient viruses from the Siberian permafrost that are 27,000 to 50,000 years old. Fortunately, these "zombie viruses" are not infectious to human beings—for now. But scientists are engaging in risky gain-of-function research that could spread these viruses to humans. Moreover, we have no way of knowing what other potential dangers might lie waiting beneath the melting glaciers.

Indeed, it might be the revival of a more recently eradicated virus that causes the next pandemic. Victims of the smallpox virus were mummified and buried—but they can easily be found, they can be dug up, and the virus that killed them can be sequenced. We've done this before to understand the Spanish flu virus.

Or the next pandemic might come from something new, created in a lab using cutting-edge scientific techniques to combine segments of DNA from different sources to create a new genetic entity. It might even come from the so far unexplored depths of our oceans. Or it might be carried to earth on a rock that NASA brings back from outer space.

But most likely this virus will come from birds. In fact, bird flu already exists. It's deadly, killing an estimated 50 percent of infected people. There already have been limited outbreaks. In January 2025 a Louisiana man who kept chickens in his backyard became the first American to die from the H5N1 virus. By that time, it was estimated bird flu had killed about 500 people around the world.

Those numbers remain low because the virus has not yet found a way to move from one person to another. Those people who have died were infected by birds. There is no human-to-human transmission.

Yet.

But it is coming. The virus is continually evolving, continually probing, searching for the slightest genetic error that will let it take hold and multiply and kill.

And when that happens, when piles of bodies are again being stored in refrigerator trucks, angry Americans are going to wonder who to blame? Whose fault is it that—once again—we were not prepared?

Here's the answer to that question: You. All of you.

The most surprising lesson learned from the Covid pandemic is how little we have learned from the Covid pandemic. We were not prepared for that and the result was devastating: The global economy was essentially shut down as we buried millions of victims. The physical, mental, and economic toll was almost incalculable. But rather than figuring out what mistakes were made, why we weren't ready, what worked or didn't work, and what we should be doing right now to make sure we don't repeat it, we essentially are doing nothing.

In fact, as incredible as it appears, we actually are less ready to respond to a bird flu pandemic than we were with Covid. I've spent my entire professional career studying, treating, and researching viral diseases. I've served on the front lines of the AIDS war and the Covid pandemic. I've been in the White House meetings and helped develop policy. And I can write without hesitation that the damage done to the nation's healthcare system by the pandemic has made us significantly more vulnerable.

The fact that we didn't know how to respond to Covid did not stop us from responding. Masks. Social distancing. Mandates. Closing schools and businesses. Restricting travel. Forcing vaccinations by enforcing penalties. We changed every aspect of life in this country, too often without scientific or historical data to support those decisions. As I explain, we did the best we could with what we knew, but in taking steps without scientific justification we made mistakes. I made mistakes. But without question the biggest mistake we made was not admitting what we knew and, probably more importantly, what we didn't know.

If we can't admit we were wrong, we are preparing the ground to be wrong again.

We were wrong to mandate vaccination. We were wrong to grant pharmaceutical companies immunity from prosecution. We were wrong to overlook the side effects of the vaccines. This lack of transparency destroyed the government's credibility as well as the reputations of some dedicated public servants. Of all the serious damage done by our response to Covid it is my belief that the most dangerous and the longest lasting is the loss of confidence and trust in the healthcare system. It is going to take a long time and a sustained effort to regain that—if that is even possible. During a healthcare crisis, it is imperative that leaders speak with a single voice. We didn't have that. An already politically divided country was driven even further apart. It is vital that any directives given be based on facts and when those facts change—as they did suddenly, unexpectedly and dramatically during Covid—we need to be open and honest about it.

We were wrong: Covid was not simply a variation of SARS as we originally believed but rather an entirely new virus unlike anything we had seen before.

We were wrong: While masks are important tools that successfully reduce transmission of the virus, they do not prevent infection.

We were wrong: There is limited evidence that closing schools prevented the spread of the virus, while there appears to be some data indicating the lost year resulted in developmental damage.

We were wrong: Although the vaccine was a historic success and saved countless lives, it did not prevent infection and offered protection for only a limited period of time. Then booster shots became necessary.

We were wrong: All Americans were not at equal risk.

We were wrong: We were not forthcoming with information as we gathered it.

As I detail in these pages, there were additional mistakes—but none of them were intentional. In only a few instances did anyone knowingly mislead the public, and even then their motives were aimed at saving lives. In the midst of a crisis, you react.

There are still many things about the COVID-19 virus that we don't know for certain. The research continues. Although there is a growing consensus the virus was created as a result of gain-of-function experiments in Wuhan, China, we don't know that for certain. But we do know that in laboratories around the world virologists are manipulating genetic material, secure in their arrogance that the results of their work will never escape.

We also know that bioterrorism remains a threat that is perhaps even more dangerous than nuclear weapons.

The next pandemic is coming. There is nothing we can do to stop it. But we can mitigate the toll if we begin preparing for it. That begins with a commitment to what I refer to as radical transparency. Get out all the information. Educate people.

That begins here. Now.

.

That dot isn't a typo. It isn't a smudge.

It's a warning.

It's small, isn't it? So small you can barely see it. Imagine something this small changing history. In fact, it's actually smaller, much smaller, thousands of times smaller. These pathogens—viruses, bacteria, fungi, and parasites—are microorganisms far too small to be seen with the naked eye, but powerful enough to cause disease, epidemics, and pandemics. They are more dangerous than all the great armies, capable of disrupting the world and leading to millions of deaths.

Even yours.

I've seen the chaos they can cause. I've seen young men dying of Ebola in rudimentary African tent hospitals. I was on the front lines of research during the AIDS epidemic and made the controversial discovery that HIV could be sexually transmitted by heterosexuals. And I was director of the US Centers for Disease Control and Prevention when the COVID-19 virus came out of a Chinese laboratory and killed more than a million Americans.

I've spent my career researching pathogens: first in the military, then in academia, and, eventually, in the government. I have seen the great variety of them: influenza, measles, hepatitis, HIV, SARS, MERS, RSV, shingles, COVID-19, and the myriad of others and their variants. I've studied them, I've experimented with them, and I've seen how they work, invading healthy cells and inflicting harm. I've watched the continually unfolding battle between pathogens and the vaccines and antivirals created to contain them. While other people refer to these pathogens as "the enemy," I don't. In the past I usually referred to them as "my colleagues." But that changed when COVID-19 emerged. COVID-19 was different. It changed everything. We had never seen anything like it. That virus became my enemy.

The pandemic began for me on December 31, 2019. It had begun as a day of celebration. My entire family had gathered for the holidays in a large house I had rented in Deep Creek, Maryland. It was the first time in years the whole family had been together. I had been appointed director of the CDC by President Trump in March 2018. My three objectives when I accepted the job were to accelerate the application of science to improve American health outcomes, to promote vaccines, which I believe are the most important gift science has given to modern medicine, and to prepare—to overprepare—this country for the pandemic that I felt certain was going to happen eventually.

I'd actually had a premonition that we were going to be struck by a pandemic. As a virologist, as a student of medical history, this possibility was always on my mind. I was always looking over the next horizon for any indication it was coming. Everything I'd learned in my career made it highly likely to me that we were due. It was as if I was looking at an equation: All the factors were there, it just hadn't been summed up yet. It was a matter of someone finishing the calculation My mistake was believing it would be a bird flu.

It had been a beautiful day. We were getting ready to celebrate the new year. My wife Joy and I were with our kids and grandchildren when my cell phone rang. That wasn't unusual; as director of the CDC, the federal agency tasked with keeping Americans safe and healthy, there was no such thing as a vacation day. There was always something going on that had to be addressed. I glanced at the number. My office was calling from our headquarters in Atlanta. I walked into a bedroom to take the call. Within seconds I learned our representatives in China also were on the call. In fact, they had initiated it.

That was unusual.

Our team in China told me matter-of-factly that there were twenty-seven cases of an unexplained pulmonary illness in the city of Wuhan, China. I knew about Wuhan; it is a large city with a population of more than 12 million residents, but it also is the home of the Wuhan Institute of Virology, a respected research center. In fact, China's first level 4 biosafety laboratory had opened there a year earlier. But I also knew that the Wuhan lab had a military component, similar to our army facility located at Fort Detrick, Maryland. The problem with laboratories, even those with excellent security measures, is they can leak. Four months earlier I had ordered the lab at Fort Detrick temporarily closed because of a biosecurity concern. In that instance I was worried that contaminated material might have gotten into the drainage system, which could have impacted the local community. That lab stayed closed for six months as the necessary biosecurity measures were strengthened.

In 2003 severe acute respiratory syndrome, SARS, a disease that was found in cave-dwelling bats, had come from a different region of China. Fortunately, we had been able to contain that outbreak. Cluster events were not unusual; ten people who got salmonella might have eaten at the same restaurant. But this concerned me. If I had been asked to predict where the next respiratory pandemic would come from, I would have said China, China, and probably China. According to my team, the twenty-seven patients had only one thing in common. They had all been to the wet market in Wuhan where live animals are sold for food.

I asked a lot of questions, but we had very little information. Among these cases, I was informed, there were three instances of multiple infections within a family. That really troubled me. The probability that different members of a family getting the same pathogen from walking through a market was remarkably small. It was far more likely that whatever this was had been transmitted from one person to another, making it an infectious disease that was transmissible to humans.

I could hear my family in the living room, preparing to celebrate. I took a long deep breath. I suspected what was coming.

In some way I had been preparing for this my entire life. I am a child of science. My father and mother were both scientists at the National Institutes of Health. My father was a medical doctor and biochemist whose dream was to figure out why my eyes are blue and someone else's are green. He wanted to crack the genetic code. In the 1940s and 1950s that was an impossible dream. While other researchers were focused on proteins, he was intrigued by nucleic acids, especially this strange thing nobody knew too much about called ribonucleic acid, or RNA. Three people from his team individually won the Nobel Prize in medicine. I've always believed he would have been among them, but when I was five years old, he died.

My mother was also a biochemist. Eventually she left NIH to work with a brilliant scientist named Herb Weissbach at the Roche Institute of Molecular Biology in Nutley, New Jersey.

My mother instilled in me a passionate love for science. I was brought up to believe in God and to trust in science. We actually lived very close to the NIH. After school my brother, and later my sister, and I would often hang out in the NIH library waiting for her to finish her work. I was about eight years old the first time I looked through a microscope. We lived in a small house. Our basement was divided by a staircase into my bedroom on one side and a small laboratory, with a basic microscope, on the other side. I remember pricking my finger and looking at my blood under that microscope. I was hooked.

My plan was to become a genetic surgeon. I wanted to be just like my father, who was a pioneer in human genetics. Rather than curing or saving patients by taking off arms or legs, I believed I could do that by taking out or putting in genes. I could do that, I believed, by manipulating viruses, which would transport those genes.

Viruses have always intrigued me. And baffled me. A virus is a microbe. It's one of the building blocks of life. It contains a strip of either nucleic acid, either DNA or RNA, surrounded by protein. Viruses are found in almost every living entity, including plants, animals and, of course, human beings. A virus can't replicate outside a cell. It survives by attaching itself to the surface of a cell. Then it digs into the cell to use its components to essentially reproduce. Too often, that process kills the host cell, causing disease and, in some instances, death.

As a teenager I came to suspect that viruses might cause human cancer. That's what I wanted to investigate. Dr. Herb Weisbach got me a summer job working in the lab of Dr. Saul Spiegelman at Columbia. Spiegelman, a famed microbiologist who should have won a Nobel for his groundbreaking RNA research, was studying viruses in cancer cells. I worked there for two summers. Ironically, as it would turn out, he actually was studying retroviruses, tricky little pathogens able to convert RNA into DNA in a cell, which at that time no one believed existed in human beings. But, as we discovered, retroviruses turned out to be the cause of AIDS.

It was while working at Columbia I began to understand that science is the art of pulling data and information out of nature and applying it to the human condition. It fulfilled my passion. It was endlessly fascinating, challenging, and mostly fun. I learned that science evolves, and that sometimes it becomes necessary to reevaluate your conclusions. I learned the importance of observing, of listening, and of being willing to admit publicly that, based on the data, your original conclusion was wrong.

What I did not learn, at least not there, not yet, were the politics and pressures of science and the dangers of scientific arrogance. All that, unfortunately, would come later.

I went to Georgetown University on a military scholarship. I intended to become a gynecologist, delivering beautiful, healthy babies while continuing my research on the possible relationship between viruses and cervical and breast cancer. My mother was making $11,000 a year. That scholarship was the only way I could afford college. But it required me to serve four years on active duty after graduation. The day I graduated I also received my "Greetings from the President of the United States" letter. I was ordered to report to Fort Bragg for basic training, and then I was going to Vietnam.

I had already been accepted into Georgetown Medical School, so I tried to defer my active duty, but the army turned me down. Every week during basic I reapplied, and every week my request was rejected. Suddenly, though, President Nixon announced we no longer would be sending troops to Vietnam. In return for extending my enlistment, the army agreed to pay for my medical education at Georgetown Medical School.

During the summer of my junior year in medical school, I met Joy Hoke, a nursing student working at Sibley Hospital. Joy was planning to become a midwife. We went out for the first time in September 1975. Six weeks later we got engaged. In January 1976, we got married. We had our future planned. I was going to be an ob-gyn and she was going to deliver babies.

That was half a century, six children, fourteen grandchildren, at least one epidemic, and one worldwide pandemic ago. And a family tragedy.

A year later Joy gave birth to our first child in Georgetown Hospital. Tragically, our son John Paul died of significant obstetrical complications. That had a huge impact on us. The pain of that could have destroyed our marriage; instead, it bound us together even more tightly. But it also changed my career path. The obstetrician who was responsible for our son's death came out to the waiting room and told me I actually owed him a debt of gratitude. I was stunned by that. He explained that he had used his skill to ensure our child was born alive, which made our medical expenses tax deductible. A tax deduction. Fortunately, I didn't hit him. Later, we learned this doctor had been impaired. Unfortunately, another baby had to die before he was stopped.

We ended up suing Georgetown for negligence while I was still a student there, an odd situation, eventually accepting a settlement.

That experience changed me. I did some of my training in neonatology, high-risk births, at the Tripler Army Medical Center in Hawaii. I was part of a team that specialized in difficult births. It was much harder than I imagined—not the medicine itself, but everything that surrounded it. I found myself standing in a hallway speaking to a young father, trying to explain the situation. "I know what you're going through," I'd tell them.

They doubted that. How can you possibly know that? they asked.

I explained. It was incredibly difficult. After having done that several times, I knew I couldn't function as a doctor. I never again wanted to put myself in the position of having to tell a young man he had just lost his baby or, even worse, his wife. I just couldn't do that again.

I still believed I could be an oncologist. But in the army that meant being stationed at a smaller base, like Fort Bragg or Fort Leonard Wood, places that did not have research facilities. There was only one specialty in the army that led to laboratory research: infectious diseases. That was my path to viruses. Rather than applying for an oncology fellowship, I decided to apply for assignment to the prestigious Walter Reed National Military Medical Center and Walter Reed Army Institute of Research in Washington, DC.

At the turn of the twentieth century, army Major Walter Reed's research proved that yellow fever was transmitted by mosquitoes, and the facility named after him became one of the world's leading research centers. The military was well aware that throughout history more service members had been sickened, disabled, or killed by infectious diseases than had been killed or wounded in combat. The military created Walter Reed to develop treatments and vaccines necessary to keep troops healthy.

At that time Walter Reed only accepted two people each year to its Infectious Disease Fellowship Program. One of those slots was filled by applicants working in the field. That left only one position for people like me, young officers completing their internal medicine residency. The choice came down to me or my best friend at the time, a West Point graduate who also wanted to study infectious diseases. One of us would get it, the other one would be assigned to Korea and could reapply in two years. We were pretty much evenly qualified. The Army's Surgeon General Consultant for Infectious Diseases couldn't make up his mind who to select. So we agreed he would flip a coin. A nickel. "Tails," I picked.

Tails. I still have that coin. It changed my life.

My education into the world of infectious diseases began the day I walked into Walter Reed. It has never ended. By definition, an infectious disease is a disorder or an illness caused by microscopic organisms. Smaller than that dot at the opening of this chapter. They exist pretty much everywhere in nature and are unavoidable. Some of them can be spread to human beings in a variety of ways: direct contact with an infected person, inhaling airborne particles, insect or animal bites, eating or drinking contaminated food or water, or even from touching surfaces. In most instances they're harmless. Our immune system protects us from many of them. Good hygiene, antibiotics, and vaccines developed at places like Walter Reed offer additional protection.

In most cases. But pathogens can and do cause diseases ranging from the common cold or flu to deadly outbreaks. The Black Death, the bubonic plague caused by bacteria spread by fleas, ravaged Europe in the 1300s, killing as much as half the entire population. London's Great Plague of 1665 killed tens of thousands of people. The so-called "Spanish flu" of 1917 was a viral disease that infected as many as 500 million people and killed an estimated 4 percent of the world population.

Since then, there have been countless epidemics that spread wildly and killed randomly. Fortunately, most of them have been somewhat contained. But we went to work every day knowing what was possible. Eventually, we knew, we feared, there was going to be another worldwide pandemic. The world had become an incubator, an almost perfect environment for a pathogen to emerge and spread. With a continually expanding world population, with people regularly traveling great distances, with cities more densely crowded than ever in history, and with tremendous poverty, it was inevitable. At Walter Reed in the 1980s, we were on the front lines of infectious disease research and prevention. We were preparing for the fight that was coming. It was an incredible place to work.

At Walter Reed, I had the opportunity to study the pathogens affecting people around the world. Influenza. Measles. Dengue Fever. Japanese encephalitis. Hepatitis B and hepatitis C. Ebola. Diseases most people had never heard of and did not even know existed.

My first real investigative interest at Walter Reed was hepatitis B. Hepatitis B is an inflammation of the liver that can lead to serious, in rare cases even fatal complications. It can be caused by several different viruses as well as alcohol and drug abuse. It was believed to be transmitted primarily by exposure to infected blood or other bodily fluids. It generally was accepted that our soldiers only got infected if they used IV drugs. I had set up the hepatitis clinic at Walter Reed, taking care of infected soldiers when they came home. Over time, I noticed I was also seeing spouses, mostly wives, but also some husbands with hepatitis B. I gradually came to believe at least certain strains of the disease could be sexually transmitted far more often than previously believed. In those days, although we had an effective vaccine to prevent it, we didn't have an effective treatment once you were infected. So I had to warn chronic carriers that they risked infecting their spouse. These were all young, sexually active men. I was giving them an impossible choice.

I decided to find out how easily hepatitis could be sexually transmitted. This was a terrific learning opportunity, the application of basic science, and it would be the foundation of almost everything I did later in my career. My hypothesis was that hepatitis could be sexually transmitted; I intended to use data and laboratory test results to either confirm or deny it. Coincidentally, my mentor had developed and was testing a vaccine for gonorrhea, a sexually transmitted disease caused by bacteria running rampant among our troops in Korea. As many as 30 percent of our soldiers were infected. As part of his trial, he was circulating a questionnaire and allowed me to add a few questions: Have you ever had hepatitis? How long have you been in Korea? In addition, I requested a blood sample.

In science, the numbers tell the story. There is no getting around that. You can hope, you can wish, you can have faith, you can try anything you'd like—but the numbers are going to give you the answer. That was as true when I was starting my career at Walter Reed as it was when I was running the most sophisticated disease tracking organization in the world. The numbers don't lie. By screening for the presence of antibodies, it is possible to determine if someone has had hepatitis B. It turned out that about 3 percent of soldiers had hepatitis B before arriving in Korea. But more than 30 percent of those troops who had been in that country for a year tested positive, which meant a lot of our people were becoming infected. Since very few soldiers had a spouse there, the disease apparently was being spread by "hostesses," commercial sex workers.

The practice of good science does take you to some unexpected places. Truthfully, during my training I never anticipated conducting a study of Korean prostitutes. We evaluated a group of these women; between 10 and 12 percent tested positive for the antigen, meaning they were highly infectious. We had our source. I pretty much proved that soldiers were getting hepatitis B by assignment to Korea. Fortunately, I knew, there were vaccines that could prevent infection.

I presented all the data I had compiled to the Armed Forces Epidemiology Board, recommending that soldiers be vaccinated before coming to Korea. It was a simple way of solving the problem, and maybe we would save some lives. To my surprise, my request was denied. It was too expensive, I was informed; the cost was something like $100 a shot. It would cost millions of dollars to vaccinate all the troops in Korea.

I refused to give up. I wondered: Was there a less expensive way to provide the vital protection? Vaccines work by stimulating an individual's immune system to produce antigens, a substance that produces antibodies or other molecules that attack the pathogen. But there was no good data indicating what dose of the hepatitis vaccine was required to produce a sufficient volume of antigens. So I started experimenting.

I designed a clinical trial, comparing my volunteers who were given a standard dose of the intramuscular vaccine to the efficacy of as little as one-tenth of the standard dose. In addition to that, I had another idea I wanted to try. In practice the vaccine was given intramuscularly, a shot right into muscle tissue. But two decades earlier, a study had proven that antigens given intradermally into the skin were as good, or even more efficient, than shots into muscles. There were people who told me that wouldn't work, it was too difficult, but I pointed out that the rabies vaccine was administered intradermally—and that vaccine saved lives.

The results of my trial were conclusive. One-tenth of the recommended dose given intradermally was as effective as a full dose given intramuscularly.

I compiled my evidence and presented it to the board. I showed them we could reduce the cost by 90 percent while providing complete protection. We could stop it. The board approved my suggestion. All soldiers going not just to Korea, but eventually anywhere in the world where they might have sexual relationships with high-risk individuals, received the vaccination. Within six months, my clinic at Walter Reed was closed because we didn't have anyone coming home from Korea with hepatitis.

I think that is one of the most important things I did during my military career. But what it really did was set the stage for my next, and far more controversial, battle with bureaucracy. A battle that prepared me, at least a little bit, for the extraordinary events and public pressure I would encounter during the COVID-19 pandemic. Until then I had not been exposed to the politics of public medicine. My data had been pretty straightforward; it was what it was, no one had an agenda. That was about to change.

And it almost cost me my career.

My hepatitis experience had given me confidence. I learned that I could come up with a new idea, stay focused on it, use the scientific method to either prove it worked or it did not, and I could persevere when criticized.

A fundamental question faced by scientists, researchers, and many doctors is how much information should be communicated to the public or to a patient. And when. Information, like elections, has consequences. It can have a significant economic impact. It may change lives. In Henrik Ibsen's classic play *An Enemy of the People*, for example, the medical officer in a small Norwegian town discovers that the water flowing into the town's public baths is contaminated and poses a significant risk to bathers. But the town's economy and reputation are dependent on visitors enjoying those waters.

The town's government, which includes the medical officer's brother, warns him that publicly releasing this information will destroy the town. They put tremendous pressure on him, threatening him with the loss of his job and his standing in the community. But finally, he decides to reveal the potentially lifesaving truth, and by doing so becomes an outcast, an enemy of the people.

In the mid-1970s, novelist Peter Benchley and director Steven Spielberg posed a similar dilemma in the novel and movie *Jaws*. In this story, a summer resort community is threatened by the presence of a great white shark that has been attacking swimmers, but keeping vacationers safely out of the water may destroy the summer economy. In this instance, the mayor decides not to issue a warning. As a result, people die.

Only a few years after *Jaws* had terrified people around the world, I found myself in the same kind of moral morass. After finishing my fellowship in infectious diseases, I decided to stay at Walter Reed and was assigned to the Department of Virology to be fully trained as a virologist. It was a great fit for me. While conducting my hepatitis research, I had started seeing a trickle of patients showing a variety of unusual symptoms. I had no idea what was causing them. A virus? A bacterial infection? Other physicians were reporting the same thing. There was nothing in the medical literature that answered my questions. We had never seen anything like this. The infection, whatever it was, did not respond to any treatment.

In 1981, we were just beginning to recognize this new infectious disease that, for lack of a better name, we called AIDS, acquired immune deficiency syndrome. It expressed itself with a variety of different symptoms, among them infections, pneumonia, and rare cancers. We knew almost nothing about it, where it came from, what caused it, how it spread, how to treat it, and how to stop it from spreading.

A few years later, I would publish an article comparing this period to the way society had dealt with the last mysterious epidemic: syphilis. For centuries after it was recognized, there was little research into its cause or treatment. When it first appeared at the end of the fifteenth century, respected scientists speculated it was caused by movements of the stars and the moon. And, as I wrote, "Society responded with social and legal interventions based on mystery. Scientific knowledge of etiology, methods of transmission and pathogenesis remained unavailable for centuries." Instead, this plague was simply accepted as part of life, and nothing could be done about it.

That's about where we were when I starting to care for my first few AIDS patients. We had no effective treatment, so there wasn't very much I could do to help them other than treat the symptoms, but I was fascinated by this new potential plague. All of a sudden, I was watching beautiful young men and women dying, despite everything I did. Reflecting back on that time, I believe the experience of watching my son die allowed me to have a much more impactful relationship with my patients. Historically, I had followed a traditional medical path; if a patient was sick and dying, I retreated. But after John Paul's death, I learned not to retreat. That became very important, as I found myself caring for literally hundreds of men and women who were going to die.

I went to work every single morning hoping someone, somewhere had made some progress in fighting this thing. That maybe we were going to have some real tools that we could use. In 1983, a laboratory in Chicago believed they had discovered the cause of AIDS. They submitted a paper to *Science* magazine claiming it was caused by a tick-borne bacterial organism named ehrlichiosis, even though that organism had never been reported to be harmful to humans. The editor of that magazine was uncomfortable publishing it but didn't want to reject it. Instead, he created a task force to go to Chicago and review the data.

The team was hand-picked by the world's expert on ehrlichiosis, veterinarian Dr. David Huxall, who was then conducting research at Fort Detrick. He had first encountered the pathogen when dogs that had been used to crawl through booby-trapped Viet Cong tunnels in Vietnam became infected. Huxall asked my commanding officer, Colonel Phil Russell, "Who do you have who's not doing anything important?"

That's how I ended up assigned to this team, which consisted of seven prestigious virologists, among them the scientist who discovered the Lyme bacteria, and me. Huxall taught me everything that was known about ehrlichiosis in about a day. Our assignment was to look at everything, to gather as much information as possible, to try to determine if this Chicago group was right.

Among the people I began working with was a man named Tony Fauci. By that time, he was already among the most respected people in the field, and he was running the National Institute of Allergy and Infectious Diseases' Laboratory of Immunoregulation. That was pretty exciting for me; I was just starting my career, and I was working someone with that experience and achievement.

It was the beginning of a relationship that would span four decades.

Tony Fauci and I both got blood and tissue samples from people suffering from this disease. I got to study them to figure out if ehrlichiosis was in fact the causative agent. Over the following months I became fascinated by this thing called AIDS. It was unique, different. No one had ever seen anything like it. I was hooked. It was the kind of scientific mystery I'd dreamed of solving when I was looking through a microscope in my basement.

The army was sending all of its AIDS patients to Walter Reed. I got to study them, I got to know a lot of them, and I became close friends with several of them. Eventually, they all died a horrible death. Every one of them.

That was my real introduction to the power, the strength, and the incredible dangers posed by these pathogens. It was only the beginning. Death was to become part of my life

Our team eventually concluded that ehrlichiosis did not cause AIDS. But we still didn't know what did. Tony Fauci's group isolated a fungus and for a brief time suspected that might be involved, but it turned out to be a contamination. I suspected the disease might be caused by a retrovirus, so I went over to the National Cancer Institute and met with the director of the Laboratory of Tumor Cell Biology, Robert Gallo. Gallo was acknowledged as one of the world's leading scientists; he had just won the first of his two Lasker Awards, America's most prestigious scientific honor, "for his pioneering studies that led to the discovery of the first human RNA tumor virus [the old name for retroviruses] and its association with certain leukemias and lymphomas."

In other words, Bob Gallo had proven that retroviruses actually caused certain forms of cancer. Now he was searching for the cause of AIDS. This was cutting-edge science. I wanted to be part of that research, and I could be valuable. In my clinic I had been identifying patients who had this disease before they could be diagnosed with AIDS. I had compiled real-world clinical material that few others could provide. I was beginning to find evidence that rather than being an infectious disease that someone acquired and within a few days became symptomatic, like a cold or flu, this was a progressive disease and might have been in their system and growing for a substantial period of time. Maybe even for years.

Bob Gallo welcomed my participation. There are scientists and researchers who are very protective of credit. I understand that. There is tremendous competition for the grants and other funds that finance research. Getting credit for a discovery and having your name included on an important scientific paper can make a significant difference in funding. Gallo was never one of those people. He appreciated the fact that I was one of the first doctors to try to treat AIDS patients. As it turned out, I had been on the front lines even before we knew the war had begun.

In 1984, Bob Gallo and French scientist Luc Montagnier at the Pasteur Institute in Paris independently discovered that a retrovirus, the human immunodeficiency viruses, HIV, was the cause of the disease. That same year, Gallo published four papers in *Science* announcing his discovery. Several years later, the question of what each man had contributed to that discovery would result in a still unsettled controversy. That never affected me. Because of my contributions to our research, Bob Gallo was kind enough to include me on the first paper he published. That was a big, big deal for my career.

I remember the first time I actually saw HIV. It was a picture taken by an electron microscope. This was before Gallo's first paper was published, so this was not public information. I was sitting in my office, and I just stared at those photos. It was stunning, like seeing a planet or a star for the first time. This was a retrovirus. This is what it looked like.

That was only the beginning of my collaboration with Gallo. A decade later I joined him and his longtime colleague Dr. William Blattner in founding the Institute of Human Virology at the University of Maryland. This was the first center that combined patient care, research, and prevention programs at one site. In some ways, it was similar to what I had been doing at Walter Reed. Our objective was to further understand how viruses worked, to identify threats, and to work to develop effective treatments.

Identifying HIV was a gigantic first step in understanding AIDS. But we still knew almost nothing about the disease. We weren't even certain how it was spread. During that period a resident at Walter Reed whom I had helped train, Bryan Raybuck, told me his wife, a physician named Deborah Brix, was having a difficult pregnancy. Her hematocrit test, a measure of red blood cells, was very low. Normally, that wouldn't have been a big deal. She simply would receive a transfusion which would solve the issue. But this wasn't a normal time. A lot of people suspected that HIV, like hepatitis, could be spread through tainted blood. Bryan asked me if she should be transfused.

I told him that several years earlier my wife, Joy, had suffered a miscarriage and also had a very low hematocrit score. Rather than a transfusion, we carefully monitored her, and eventually her blood pressure stabilized. Bryan and Deborah also decided to wait, which turned out to be a very good thing. Deborah had a rare blood type. That blood she would have received was instead given to children in the neonatology unit. Unfortunately, most of those children were infected with HIV. Unfortunately.

But that was my introduction to Dr. Deborah Brix. She had done her fellowship working with Tony Fauci, but soon after this situation I asked her to come work with me. Beginning in 1985, she worked as my deputy for more than a decade. We became close friends. Certainly, none of us, not Tony, not Deborah, and definitely not me ever considered the possibility that decades later the three of us would come together to lead the fight against the first pandemic in more than a century.

Meanwhile, I had become the army's leading researcher into this new virus. There wasn't much competition for that role. Because we knew so little about how it was spread a lot of people were reluctant to treat AIDS patients, "reluctant" being a nice way of saying they wanted nothing to do with these people. Some of my colleagues warned me about getting too deeply involved with this disease. The general consensus at the time was that AIDS was mostly limited to homosexuals and drug abusers, populations that didn't fit the image the army wanted to project. Don't get too closely associated with this, people I trusted whispered to me. It wasn't just the physical danger, this was so controversial it potentially could end my military career.

Admittedly, I thought about that. Joy and I discussed it. We had little kids, and no one knew what I might be bringing home. But there was never any doubt that I would continue with my work. More than anything, I think my decision came down to what I had learned from my mother: Science should be used to improve the human health condition. In some ways I felt like a detective at the beginning of a complex murder case. I had a growing number of bodies, and now I knew the weapon: HIV. All I had to do was figure out everything else.

While I was arrogant enough to believe I could contribute to the science, I had a pretty strong suspicion this was going to be a lot more complicated, a lot more dangerous than most people believed.

Slowly, painstakingly slowly, as this plague began devastating the gay community, we began to learn more about it. Initially, it was believed, this virus was transmitted through infected blood, most often by anal sex or using dirty needles. Little was known about the incubation period. Those beliefs were comforting to heterosexuals—other than rare instances, primarily blood transfusions, this wasn't going to affect them—so there was considerable interest in maintaining that fiction.

I don't know exactly when I suspected that wasn't true. I remember flying to Denver in 1984 to meet with the army's top virologist. I wanted to show him the electron micrograph which was not yet in the public domain. Here it is, I told him. This little virus is the cause of AIDS, and we know almost nothing about it. I was trying to lobby for the additional funds that would be required to do the research.

That was the first of many meetings I would have as I tried to spread the alarm.

This was going to be a big deal, I felt that very strongly. I remember going to a library to learn about the infectious diseases that had influenced the balance of power in the world. There had only been a few of them. I was sort of surprised there was not more information about the subject. As I sat in the library, it felt like a storm was coming, but there was little to no precedent for what it would look like.

I felt the same uneasiness, the same inevitability about it then as I would later come to believe about bird flu.

TWO

When I was named director of the CDC in 2018, my goal was to transform the agency into the tip of the spear. I was quite certain we would face significant challenges. If I were a betting man, I would have put my money on bird flu. But whatever it was, I intended to prepare for it. The proven way to stop the spread of an infectious disease is to detect, respond, and prevent its spread. Early detection is essential. If we can learn about a potential outbreak while it is still contained, we can isolate it. To do that, I wanted to build a network of bases around the world. These bases would serve as our eyes and ears, our frontline response to pathogens and chemical warfare. We actually began that effort.

We had three Ebola outbreaks during my tenure. Ebola is a severe and often fatal viral disease usually found in Africa. These cases were in the Democratic Republic of the Congo. Ebola is a hemorrhagic disease. It causes massive bleeding, and it spreads through contact with bodily fluids or contaminated objects. It does have the potential to become a worldwide epidemic, but it can be contained. We caught the first outbreak early because we had a programmer there. Working with the DRC's Ministry of Health we stopped it after about fifty cases.

The second case took place in a war zone so it was far more difficult to get our people there. We also detected it very early and could have stopped it, but the State Department prevented us from going there at the beginning.

Eventually, we had an experimental vaccine. We tested it with a "ring" vaccination, giving the vaccine to anyone around the patient. When the next case emerged, we did the same thing. It turned out that the vaccine was reasonably effective. It helped stop the spread of Ebola, and eventually the epidemic in the DRC was brought under control.

I had learned the importance of that early warning system in those first harrowing days of the AIDS epidemic, when politics and fear made it difficult to properly inform the public. But as detrimental as that was, that was only my small introduction to what was going to happen during the Covid pandemic.

By 1982, it had been firmly established in public perception that AIDS almost exclusively affected gay men and intravenous drug abusers. But that was not what I was seeing in my clinic. As many as 30 percent of my forty-five patients were women, and there was little evidence that all of them were shooting drugs. Even more puzzling, 50 percent of my patients were married, which gave me the chance to evaluate spouses. It was obvious that husbands and wives were infecting each other. Eleven of the twenty-seven women married to men with AIDS and four of the nine men married to women with AIDS became infected. I was looking at evidence that the general perception was wrong, that in reality AIDS was bidirectionally transmitted. That meant heterosexuals also were at risk of being infected with this deadly virus.

I can remember the instant I realized that was true. Literally, the instant. That's the impact it had on me. I had been studying seven couples. Either the husband or wife had been diagnosed with AIDS but their spouse had not. I had sent viral cultures from them over to Bob Gallo's lab at NIH to be analyzed. On a Sunday morning,

I was at home reviewing the data. The results showed that five of those seven couples had infected each other. I just sat there looking at the data. There was no other explanation for the presence of low T cells, a marker for the disease.

I knew what it meant. This was definitive proof that this virus was a bidirectional heterosexually transmitted disease. I knew the possible ramifications. At that moment, I was one of very few people in the world who had this knowledge. I have to admit that my first response was not how terrible this was, but rather the excitement of discovery. But within a few minutes, that excitement turned to horror. Now I knew we were dealing with a sexually transmitted fatal disease. It was not unique to homosexual activity. I sort of assumed the reason gay men were more infected was simply because that community had more sexual exposure to HIV.

In the previous few years, we had seen how quickly and easily the sexually transmitted herpes simplex virus had spread. AIDS was capable of the same level of transmission, and it was potentially deadly.

AIDS was going to affect the entire world, I had no doubt about that, but I was thinking mostly of my children. I had four living children who, at some point in their lives, would become sexually active. And they potentially would be at risk from this plague. I became an activist that day.

I was amazed how many public health professionals just didn't want to admit this. My education into the reality of the practice of public medicine was about to begin. As the army's leading AIDS clinician and researcher, I was asked to speak at the first International AIDS conference, which was held in Brussels, Belgium. There probably were about ten thousand people from around the world there, all of them working in public health. The goal was to share whatever information we had and to try to develop some unified international response.

I was surprised by how much questionable information was presented. There were people from several African countries, for example, who insisted AIDS was not yet a serious problem in that part of the world. I understood these people were trying to protect their fragile economies, but the cost in lives was going to be horrendous.

New York City's Health Commissioner told the audience that there was no evidence of heterosexual transmission in that city. Someone asked about the validity of my study, and this man dismissed it, saying no one in the military tells the truth. He claimed that all of the infected women in New York were IV drug users.

I got up and challenged that. I asked the commissioner how many women in New York that don't use IV drugs, who may be prostitutes who use cocaine nasally, were tested? Standing in front of ten thousand people, he replied that he didn't test anybody like that. His reasoning was stunning: "Because those people can't get infected."

This was the health commissioner of a city of seven or eight million people. The dangers of that type of thinking were obvious.

The assistant secretary of defense came after me because I was contradicting the public health service. He actually claimed the reason both men and women in Africa were becoming infected was because they were eating monkey meat and using anal sex as birth control. These were intelligent, high-level people.

I also became embroiled in the debate about when people should be diagnosed with the disease. Eventually in the mind of the public HIV and AIDS came to mean the same thing: You had the disease. But they are very different. HIV is the infection. AIDS is the full-blown disease. It was not known how long the virus was present in a person's body before it became AIDS or even what percentage of infected people actually progressed to the disease. Both the CDC and the NIH were estimating that about 5 percent of infected people got sick. There was no scientific basis for that guess. I did not have any idea where that number came from because that wasn't what I was seeing in my clinic. I had been following infected people for eighteen to twenty-four months. Based on that, I had developed a six-level staging system depending on their immunodeficiency. I eventually published my results in the *New England Journal of Medicine*. It was easy to sum up. Ninety-two percent of the patients I followed got sicker within eighteen months. Eventually almost all of them developed AIDS. This clearly was a progressive disease. If you had HIV, the odds were overwhelming you would eventually develop AIDS.

When do we diagnose this disease? That became a very controversial question. The government took the most conservative position, deciding physicians should diagnose AIDS, not people who are just HIV-positive. They were beginning the diagnosis at the end clinical stage of the disease. Well, I thought that was a dangerous decision. I suppose if they were right, that in the great majority of cases the infection disappeared without causing harm, that might have made sense. Why expose people to the growing stigma if it was only a temporary condition? But that wasn't reality. That decision was based on wishes—not science.

I've always believed that knowledge is better than ignorance. Even unpleasant knowledge. The fact was that this was undoubtedly a progressive disease. If you were infected with it, I felt strongly that you had a right to that information. Maybe even more important, most people—if they knew they were infected— would not want to transmit the disease to others. Early diagnosis was a vitally important first step in stopping this epidemic. We had the science to test blood for it, and we knew who was positive. But incredibly, people who tested positive were not informed they had this disease. That was awful. The American Red Cross was testing all blood donations. But while they were discarding infected blood, initially they were not informing donors that they were positive. They felt that individuals had the right *not* to know about that diagnosis.

The policy was putting people in real danger. In several states, New York and California, for example, it was illegal to put a positive HIV test in a patient's medical records. That meant that an obstetrician could not tell the pediatrician delivering a baby that the woman giving birth had AIDS. There is a lot of bleeding associated with a birth, putting everyone participating in jeopardy. It was an insane policy.

I felt so strongly that what you didn't know—in this case, quite literally—could kill you. In addition to blocking unwitting transmission, there were a lot of medical reasons supporting early diagnosis. By drastically and severely misrepresenting the magnitude of the epidemic, we were creating a false sense that this could be contained. As a consequence, we didn't take immediate and drastic steps to fight it. Additionally, we did not serve the needs of our patients. Even though we didn't have a cure, there were treatments for certain conditions. For example, we knew how to treat tuberculosis and syphilis more appropriately in the setting of HIV.

I understood why people opposed me. There was a tremendous level of social stigma attached to the diagnosis. People were scared. They were terrified they would catch this disease. Nobody wanted to be around an infected person. A lot of doctors refused to treat HIV-positive patients. Men and women wouldn't date someone with the disease. There was a probably legitimate fear of discrimination. People feared a positive diagnosis would cause them to lose their health insurance and their life insurance. There also were a lot of ugly rumors and accusations. There were misguided people who thought it was some kind of retribution for homosexuals. We began hearing stories that within the gay community AIDS patients were intentionally infecting as many people as possible. It got very nasty very quickly.

We knew too much without knowing enough.

It was a nightmare.

As Henrik Ibsen captured in *Enemy of the People*, at times science tells us things we don't want to hear. People react to that information in various ways. Some people choose to deny it. I felt we were facing a potential global medical crisis. I believed this disease, if unchecked, could affect the global balance of power. It was potentially that dangerous. I wouldn't—indeed, I couldn't—keep my opinion to myself. It was a very unpopular position. I began my own little campaign to warn the army that this was a significant threat. When I started talking about it, I was told by a general, a very good virologist whom I really respected, that I needed to cease and desist. I was only a major, but I refused to back down. He had three options, I told him: He could fire me, he could tell his superior officers that he had conveyed their message, or he could support me.

I was already getting invitations to testify. For a time, the military did not allow me to speak because I wasn't representing the official military position. To be honest, I made a lot of mistakes in those days, but this was not one of them. I was passionate about getting people diagnosed for HIV infection rather than AIDS. I insisted that we treat this not as some plague, not as a behavioral issue, but as a medical condition that must be diagnosed—exactly the same way I would later advocate for drug use disorders.

It was incredibly frustrating. Congress was allocating millions of dollars for AIDS research and treatment, but I wanted to convince them to invest much, much, much, much more money on HIV. I felt like they were ignoring reality. They were talking about the AIDS epidemic of 1986, while I was talking about the HIV epidemic that may have begun as early as 1976. As I wrote in the *New England Journal of Medicine*, "Despite the universal availability of diagnostic testing of blood products in the United States for nearly 18 months, physicians in epicenters of this disease still are restricted. . . . The logic that resists the establishment of etiologic (a cause or origin) diagnosis in clinical medicine or public health when available is unprecedented and should not be endorsed by the medical community."

The army finally accepted my recommendations. In October 1985, they began a massive clinical program. I was the major architect of the defense department's response to the epidemic. It was totally rational and medical-based. Eventually we tested more than five million men and women. About six thousand tested positive. We also tried to find some treatments or a vaccine. Obviously, we were not successful. But that policy remained very controversial. There were people who felt strongly that testing soldiers for the presence of HIV was an invasion of their privacy. The media attacked me. When I was later nominated to become CDC director, an article resurfaced from that period. In it, the writer claimed I wanted to put people with HIV in "leprosy camps." To the contrary, my policy and my actions were precisely the opposite: I was a strong advocate for keeping those people on active duty and making sure they got the care and treatment they needed. The writer of the article, without evidence, blamed me for being responsible for a variety of truly horrific outcomes.

It was, in my opinion, an outrageous and inaccurate attack on my medical ethics. But it turned out to be little more than invaluable preparation for the sensational and often erroneous stories that would be published during the COVID-19 pandemic.

To be clear, my recommendations were not always followed. People were scared. When I learned of abuses, I tried to rectify them. I was not always successful. The balance between individual rights and public safety is sometimes hard to navigate. There are situations in which one or the other has to suffer; sometimes you do have to risk becoming "an enemy of the people." My job at that time was to ensure that the United States Army was the healthiest fighting force it could be. Ignoring the presence of this disease did not fulfill that mission.

I was telling people a lot of things that they didn't want to hear. In 1986, I was invited to speak at one of the first national AIDS conventions. I presented the results of my clinical observations and research. Contrary to the accepted statement that only 5 percent of people infected with HIV would progress to AIDS, the reality was that almost everyone infected with this virus eventually would develop the disease, and we had no way to stop that.

That made the audience uncomfortable. When I told them that we had an obligation as medical professionals to inform people who tested positive, they started booing. It was just a cascade of boos. It wasn't quite Galileo being persecuted and imprisoned in 1633, but it was really distressing. I had followed the scientific method, and I was presenting the results. These people strongly supported the conclusion that we should not be diagnosing people, only their blood. That the results should remain anonymous. That meant, for example, that the Red Cross was screening all donated blood—but initially refused to inform donors that they were positive. That was a big mistake. These people were booing science.

I continued speaking. They could drown out my words, but not my conclusions. If I didn't have the courage to stand up for what I believed was the right thing to do medically for my patients, I shouldn't have bothered to show up. What was discouraging was that this was a room filled with hundreds of public health officials. It was not a pleasant day. These people had allowed their emotions to take control of their scientific training.

The great thing about science, real science, is that it encourages questioning. It demands people raise doubts and demand proof. And if other people, working independently, can't reproduce your results, then it isn't science. We saw that in action in 1998 when the highly respected British medical journal *The Lancet* published an article claiming there was a link between the measles, mumps, and rubella vaccine (MMR) and autism. Although the article did add to the debate, the results did not "prove" the connection. The article created a furor, and parents were panicked. Other scientists and researchers began examining those claims. No one was able to confirm the published results. Within weeks, a panel of thirty-seven experts brought together by the Medical Research Council announced it found "no evidence to indicate any link between the MMR vaccine and colitis or autism in children."

Over time, additional information was published confirming this was a fraud. *The Lancet* was forced to withdraw the article, an extremely unusual action. That's the way the scientific process and scientific progress is supposed to work. The claims could not be independently supported, although too many people continued to believe the bogus claim and still do.

When I finally got the opportunity to testify in front of Congress, I insisted, "Today, AIDS is our nation's number one problem. We must recognize it as such. . . . The difficulty, as I see it, is that some Americans cannot foresee the grim reality of the next decade. They will only understand when forced to confront the human suffering in the flesh."

In closing my statement, I stated my position, which would guide my actions throughout my entire career—and especially when I became CDC director, "Public health is a responsibility of government, and cooperation to maintain public health is the responsibility of each citizen. In any infectious disease, knowledge of the infection is paramount in its control. . . . Vigorous leadership coupled with accurate education and classical public health measures will limit the spread of this deadly virus in our nation."

When asked by Senator Strom Thurmond how we can prevent or limit transmission of the disease, I responded that the most important strategy was simply knowing whether "your sexual partner was infected or not." It is pretty simple: Infectious diseases infect people. When trying to protect people and prevent the spread of an infectious disease, the first thing you have to find out is who is infected. Without that information, there was no way to guarantee public safety. "I think the use of condoms clearly reduces the risk of transmission and should be easily available and promoted, but they should be promoted not as a failsafe method to prevent AIDS transmission, but as a mechanism to reduce the risk . . . the most critical thing we need right now is national leadership, whether that is at the congressional level or the executive level."

The fight to contain AIDS was long and expensive. It took us four years from the time we first recognized the presence of this virus until we were able to develop a workable test for it. And while we have discovered treatments that allow infected individuals to lead normal and productive lives, we still have not been able to find a vaccine.

At the same time, we were fighting the disease in our hospitals, clinics, and laboratories, researchers were trying to figure out the derivation of this virus. That was monumentally important. Finding the source of this was vitally important in the ongoing war against viral diseases. Theoretically, having this information might prevent it from happening again at some point in the future. Viruses are an elusive enemy; in many ways, they are the microscopic shape-shifters of science fiction, living Transformers able to find ways to overwhelm biological defenses. Eventually, it was determined that HIV in all its variations was a zoonotic disease, meaning it came from nonhuman primates, specifically from chimpanzees in Central Africa. A "cross-species transmission." The virus apparently crossed through contact with infected blood or bodily fluids.

Zoonotic diseases are relatively well-known and include potentially fatal diseases like rabies and anthrax. In fact, there are estimates that far more than 60 percent of all infectious diseases have been spread from animals to humans. It actually is unusual for a virus to move from primate to human. Even on those rare occasions it happened, it rarely spread: A human was a dead-end host. When humans were infected, they died. And the virus died with them. So it didn't spread. But eventually this virus evolved and gained the ability not just to go from nonhuman primate to human, but also from human to human. It's all hypothetic, but I believe it found a new receptor, a new place on a cell that it could attach itself to.

Once the virus figured out how to use that receptor, the results almost were inevitable. Africa was growing and the thousands of miles of highways and roads, of available transportation, made it easy for people once stuck in a small village or town to move easily.

The generally accepted consensus is that people got this virus by eating infected chimps and then unwittingly spread it. Scientists reached this conclusion after finding a similar virus in monkeys and apes. The research traced the first known transmission of this animal virus—SIV, simian immunodeficiency virus—to a human being to the area around Kinshasa in the DCR as far back as the early 1920s. The Congo River Basin is considered a biodiversity hot spot, the home of a great variety of plants and animals. So it isn't surprising that previously unknown life-forms, like this virus, would emerge from there.

In 1986, researchers found the virus in a preserved blood sample collected in 1959 from a person living in that area. This is the first known human transmission. Sometime in the 1960s, the virus was found in Haiti, presumably brought there by a Haitian working in the DRC, and from there eventually to the United States. It surfaced in America in 1981, when physicians began diagnosing patients with several previously rare diseases, including Kaposi's sarcoma.

I don't remember precisely when I saw my first AIDS patient, but I do know it was before we realized this was going to be an epidemic of massive proportions that would impact the world. This was just a few soldiers with similar, unusual symptoms. At first, I wondered if it might just be one of those strange coincidences that pop up in medicine, but the trickle rather quickly became a stream that could not be ignored. The fight that would be the focus of my career for the next decade had begun.

That fight changed my life. The fact that I was so outspoken in my disagreement with public health officials about how to deal with this made me an easy target. I didn't hold back in my criticism of them. I wrote that "most infectious diseases become established in man long before the scientific knowledge is available to alter the otherwise eventual outcome," and "rarely has man been given the opportunity to influence whether or not a disease becomes endemic within the species." I elaborated that "we have been given that rare and treasured opportunity. . . . As opposed to the fate of fifteenth-century Europe versus syphilis, we need to incorporate knowledge of the past five centuries into our approach."

Basically, that meant admitting publicly what was already becoming well-known: This was primarily a sexually transmitted viral infection affecting men and women, no different than any other STD. Except, it was deadly. Stopping its spread meant informing the public that everyone was at risk and making testing widely available. I understood one important way of stopping or slowing the spread of this disease was changing behavior, although obviously that had to be voluntary.

As it turned out, almost everything I said eventually proved to be correct. That wasn't because I was so much smarter than anybody else, but rather because I was the voice of medical truth. I was simply following the science and urging other people to do the same. I testified in front of congressional committees several times. As a result of those testimonies, I was accused of some outrageous, despicable things. But those testimonies also made me a public figure and attracted the attention of a number of politicians. My testimonies led eventually to my involvement in government.

I have never been a political partisan. I have donated to both Democrats and Republicans. I have my personal beliefs, but I have never let them interfere with the science. And I have never let them influence my public service. I was a career military officer, eventually leaving the service as a colonel. I'm not a political activist, never have been. My core belief is that people should have access to high-quality health care. Everything I have done in my career has been done to facilitate that.

I initially got involved in government policy when the Reagan administration was creating an AIDS commission and asked for my assistance. I actually did quite a lot of work behind the scenes, and I believe I had an impact in developing a national policy. That continued through the George H. W. Bush presidency. I also provided information for the White House during the Clinton years. But the first time I was vetted for a government position took place under President George W. Bush in 2000. I was told I was being considered to head either the National Institutes of Health or the CDC.

It's a fascinating, but rough process. I was interviewed by the White House personnel office several times; but, at that level, you don't get much feedback. During my interviews for a leadership position at the NIH or the CDC, I admitted my goal would be putting science in action, making whatever we had available to the public. My biggest complaint was that the CDC, in particular, had too often been an academic organization, publishing a lot of papers, but not leading the fight in the field. Even as far back as 2000, I feared we were susceptible to a pandemic—although my fear was bird flu.

Both Tony Fauci and I were vetted, and neither one of us got the position. At least not initially. Joy and I make our decisions together, and we decided we did not want to uproot our family and move to Atlanta. Several months later the White House once again reached out to me. Two names were leaked to the media as potential CDC directors, mine and Julie Gerberding. This time Joy said if the president wanted me to take that job, she would support that decision.

The president picked Julie Gerberding.

I was quite fine with that, I was a tenured professor at the University of Maryland and filled several leadership roles. I was also deeply involved in my research and clinical work at the university, overseeing a program in which we were providing care for six thousand patients in Baltimore and Washington, as well as an estimated 1,300,000 more patients in eleven African countries and the Caribbean as part of PEPFAR, the President's Emergency Plan for AIDS Relief. Our research was making significant contributions to the understanding and treatment of AIDS. I spent time in several African countries, including Botswana and Malawi, and I saw firsthand the devastating consequences of AIDS there. I also saw the difference that President Bush's financial commitment to combating this disease was making there. We were able to increase HIV viral suppression rates in much of Africa by almost 90 percent.

We also were saving lives in our clinic at home. Progress had been made; patients whom I knew would have died a decade earlier were now leading full and productive lives. In addition, I had been drafted into the fight against the rapidly growing opioid epidemic when one of my kids became addicted to cocaine. I had been shocked when I learned we were losing as many young Americans to these highly addictive drugs as we were to guns every year; that was stunning, stunning. We were losing an entire generation. I had believed I knew as much about addiction as any medical professional; I had treated hundreds of people with heroin and other addictions in the clinic, and I had run the division of addictive medicine at a major hospital. I thought I was an expert. But until it entered my family, until my son began to suffer from a drug use disorder, I had only an academic understanding of addiction. Suddenly, I saw it from a very different angle. I had become a strong advocate for effective treatment, recommending hospitals devote 10 percent of their budgets to establish treatment programs.

I would have been quite content to have spent the rest of my career working at the university. But in 2017, Tom Price, President Trump's new secretary of health and human services, asked me if I would consider taking charge of the CDC. I had several interviews, and they didn't go well. The impression I got from the questions was that Price was mostly interested in reducing the size of the agency. They thought it was too bloated. There were seven hundred job vacancies, I was told, and they wanted to increase that number to one thousand. I didn't think that was the right direction. We needed to expand the number of people we would need to meet the coming problems.

That wasn't me, I told them, I'm a program builder, not a caretaker. I wanted to expand the agency, maybe refocus it, but not shrink it. I laid out my objectives, which did not change significantly over the next two decades. I wanted to build a national HIV program. I wanted to get rid of hepatitis C. I wanted to expand the availability of vaccinations. I wanted to build an early warning system that would greatly enhance our nation's capability in biosecurity and pandemic response.

So I was not surprised when I didn't get the job. I was a poor fit for what they wanted to accomplish. Price eventually appointed Dr. Brenda Fitzgerald, then commissioner of Georgia's Department of Health. But later that year, Price resigned and was replaced by Alex Azar, an attorney and former president of pharmaceutical giant Eli Lilly's American operations. In January 2018, once again the White House personnel office contacted me. I was asked, "are you still open to becoming CDC director?"

I said, "Well, you already have a CDC director."

Watch the news tomorrow, I was told. It previously had been disclosed that Dr. Fitzgerald had purchased stock in a cigarette company. The next day she resigned. And once again, my name was leaked as the probable nominee. Truthfully, I was excited about this opportunity. It was a chance to put all my beliefs, everything I had learned, into national practice. I believed, without doubt, that I could make a real contribution to improving the health of Americans.

The attacks on my record began almost immediately. It was disappointing, but not surprising. I became a big target. The decisions I had made early in the fight against HIV/AIDs became ammunition against my appointment. My policies, wrote the *Washington Post*, included "mandatory HIV screening for the military before effective treatments were available" and "a controversial and unsuccessful effort in Congress in 1991 to require HIV testing of health-care professionals who perform invasive procedures, after a young woman . . . contracted the virus from her dentist."

The paper also criticized my "close relationship with a conservative AIDS organization that strongly supported the vaccine."

Obviously, no one possibly could have known that within months we would face a crisis of equal or greater proportion from COVID-19. I would be forced to make those same types of decisions. People forced to wear masks? Mandatory testing? Be vaccinated? Jobs at stake? It was incredible. Those policies, which according to the *Post* had "stigmatized those who were infected and feared being fired—and losing their health insurance," would be debated and tested, praised and criticized in the media. Ultimately, they would divide the nation.

Several prominent politicians were highly critical of my nomination. Senator Patty Murray (WA), the most prominent Democrat on the Senate's Health, Education and Pensions, led the attacks, sending a public letter to the White House claiming "Dr. Redfield's lack of public health credentials and his history of controversial positions regarding the prevention and treatment of HIV/AIDs" and "his ethically and morally questionable behavior leads me to seriously question whether Dr. Redfield is qualified to be the federal government's chief advocate and spokesperson for public health"

Other criticisms surfaced. In a long career like mine, it almost is inevitable that issues will arise, that some people will be disappointed by decisions, and that mistakes will be made. I didn't mind that type of criticism. But the other criticism, I didn't like it. I didn't think it was fair. I wish I could be as sanguine as Thomas Edison, who supposedly said when his laboratory burned down in

1914, destroying a lifetime of work, "Thank goodness all our mistakes were burned up." Thomas Edison never had to deal with the vast American social media.

Ethically and morally questionable behavior? That was outrageous. My reputation was being destroyed, and there was little I could do to fight back. I had been advised by people who had been through things like this not to respond, to let the process play out. I knew, I absolutely knew that every decision I had made had been based on science. I had supported the development of a therapeutic vaccine, although ultimately it failed. I had advocated for testing, both to help prevent the spread of the disease and to make people aware of their health status. I understood that good people might disagree with my decisions, and I also accept the fact that I am capable of making mistakes. I'm an adult, I can take criticism. But "ethically and morally questionable behavior?" Never.

I was learning what was later to become a vitally important lesson: I had always been aware of the potential dangers when politics, opinion, and science collide. It's a bad mix. I had seen it during the battle against AIDS. The public health suffered. After that I had done my best to avoid it as much as possible. I had maintained a reasonably low public profile. Now, once again, I was caught in the middle of it.

Fortunately, many of the people who had worked with me rose to my defense. I had been working for a long time with the greatly respected Democratic congressman Elijah Cummings, the legendary civil rights pioneer who had represented a large area of Baltimore for more than two decades. He didn't hesitate to speak up for me, telling reporters he didn't agree with President Trump on anything but my nomination. "I am in complete agreement that Dr. Bob Redfield is the best choice to lead the CDC," he said, "he is a deeply experienced and compassionate public health physician."

Kathleen Kennedy Townsend, the former lieutenant governor of Maryland and a member of the Virology Institute's board, added, "it's terrific to have someone who has been such a caring doctor, who has really treated patients and knows what they are going through."

And Jim Curren, who had led the CDC's HIV/AIDs battle for more than a decade and was then codirector of Emory University's Center for AIDS Research, a physician I had worked with for a long time, said "he understands the public health challenges of infectious disease both at home and aboard, and will be a great advocate for the national and global public health work of the CDC."

The director of the CDC was appointed by the president, so I did not need congressional approval. That meant there were no public hearings. Although it created an easier path for me, a hearing would have given me the opportunity to make my views known to the American public. The fact is that my reputation had been damaged. In this role, that was a serious problem. For public health policies to be effective, the public needs to have confidence in the leaders creating those policies. They have to believe those policies are being implemented for only one reason: to protect the health of Americans.

For decades CDC director had been an essentially nonpolitical position. But that was changing. Not the reality, just the perception. Before I even became the director, people who previously had never heard of me had already formed an opinion about me. They knew I had been appointed by President Trump, which immediately made me suspect to half the country; then they read I had strong Christian values, which to some people meant I had to be politically conservative. Other people read attacks that also colored their opinion of me. So I began the job with a significant number of Americans believing I had a political agenda. The consequences of that were going to be greatly magnified when the pandemic began spreading.

THREE

The enemy was the mosquito.

As America went to war in early 1942, the government acknowledged: "Of all the diseases transmitted to man by insects and animals, malaria was the most serious threat to civilian and military health." The mosquito could do what the Germans and Japanese could not. It could drastically slow down the training of troops and the production of the weapons they would need. "The experience of the Public Health Service in World War I . . . showed that extraordinary measures would be needed to protect troops and the public from malaria . . . as thousands of persons, both civilian and military, were moving into malarious areas in the Southeast."

To combat that threat, in 1942 the Office of Malaria Control in War Areas (MCWA) was founded in Atlanta, Georgia. The War Department gave the new bureau a small appropriation to create mile-deep mosquito free zones around 2,200 military bases and factories in thirty-two mostly southern states and the Caribbean. One mile was the flight range of malaria-carrying mosquitoes.

As the war continued, returning veterans brought home an array of tropical diseases and parasites, among them dengue fever, yellow fever, oriental hookworm, and typhus. In response, the MCWA expanded to meet these threats. Strategies included everything from draining swamps to creating courses to enable health officials to recognize these health threats.

The program proved so successful that in July 1946 Malaria Control officially became the civilian Communicable Disease Center. It was located on one floor of a building on Peachtree Street in Atlanta and had about four hundred employees, including scientists, engineers, and entomologists. It also had two army surplus trucks, which, according to the Atlanta Constitution, were "suited for the study of epidemic conditions at the place of outbreak" and could "be converted into medical clinics." The new agency's purpose, the story reported, was "to continue investigations in a broader field of insect borne and related diseases, to prevent new types of infections from being brought into the country and to combat certain endemic infections." A mission that, in many ways has never changed.

A year later, the CDC paid Emory University $10 for fifteen acres of land to build its headquarters. It has remained the CDC's home ever since.

In 1949, the nation's first disease surveillance program, essentially an early warning system, was launched. The center began training epidemiologists, who would go wherever necessary to identify outbreaks and try to determine the cause.

Throughout the 1950s, the CDC began establishing its reputation. When children who had received the new Salk polio vaccine began becoming infected with the disease in 1955, these so-called "disease detectives" traced the problem to a contaminated batch of the vaccine being produced by a California laboratory, helping to prove the vaccine itself was safe and efficacious. Two years later, the CDC located the source of a massive flu epidemic and began guidelines for a vaccine. Among the most significant programs was working with the World Health Organization to vaccinate millions of people in a successful effort to eradicate smallpox.

In 1970, the expanding agency officially became the Center for Disease Control. By the time I was named director in 2018, the Centers for Disease Control and Prevention, as it was known, had almost 12,000 employees spread around the world and a $11.5 billion operating budget. That "and Prevention" had become a significant aspect of the agency's mission. The CDC was responsible for investigating the entire spectrum of public health challenges facing the country, including chronic diseases, disabilities, injury control, workplace hazards, environmental health threats, and bioterrorism preparedness.

The CDC had a long history of protecting the health of Americans. Its scientists and researchers discovered the cause of puzzling killers like toxic shock syndrome and Legionnaires' disease. It had compiled the data that led to laws mandating the use of automobile seat belts. It had shown that lead in gasoline was affecting children. It had traced numerous foodborne toxins to their sources: factories, farms, and restaurants. And it had been the world leader in protecting people from infectious diseases, including SARS, MERS, and Ebola.

I knew what I intended to do when I accepted the job. I had spent four decades in public health. I knew the system, what worked, and what needed to be done to improve outcomes. What I could not possibly have known is how those goals would align with the extraordinary challenges the country would face.

Among my primary objectives was to take proven science off the shelf and put it to work. I was well aware, for example, that Alexander Fleming had discovered penicillin in 1928. It came to be considered a miracle drug, the first antibiotic; but it took more than a decade, until the 1940s, before people realized its extraordinary value and it came to be considered the miracle drug. How many lives might have been changed if penicillin had been put into action in the 1930s?

I love research. I love discovery. But discoveries need to be translated into interventions to change public health. In my opinion, the

CDC had evolved into an organization more comfortable contributing to the scientific literature than applying the advances that had been made in clinical medicine and science to public health. The reality was that we had the knowledge, the proven science to cure diseases, but we weren't applying it. I knew that from my own work. Scientists had found a cure for hepatitis C in 2013; but in 2017, 18,000 people died from that disease, and as many as 80 percent of the people who had that disease still hadn't been treated. We had a proven method for saving lives. Yet, thousands of people were still dying. That made no sense, and I wanted to change it.

Ironically, as it would later turn out, I wanted to emphasize vaccines as a proven method of preventing disease. For some reason, I couldn't quite figure out why so many people were refusing to get safe, proven vaccines. We were in the middle of a measles outbreak across twenty states when I took control. That was puzzling. Measles had been eliminated in the United States in 2000. The vaccine is incredibly effective. But it doesn't work if people don't get it.

There are more than two hundred human papillomaviruses that cause a great variety of sexual transmitted diseases, including cervical cancer in women and throat cancer in men. Between 2017 and 2021, almost 50,000 Americans were diagnosed with HPV cancers. Vaccines have been developed that make it almost totally preventable. Australia, for example, has practically eliminated cervical cancer. But in the United States, less than 60 percent of the at-risk public, especially children and teenagers, had been fully vaccinated. I didn't understand it. This is like a get-out-of-jail free card for a potentially fatal disease. I felt certain that if we told people, "hey, we can vaccinate you against a terrible cancer," they would jump at it. But apparently, we were not getting that message out.

High on my list of priorities was convincing more people to get the annual flu vaccine. Influenza was probably the most significant health threat we faced. The previous winter, the flu had killed 52,000 people. Between 2008 and 2018, influenza had caused the premature death of 360,000 Americans. That was more than seven times the number of people who had lost their lives in Vietnam. How could that happen?

This effort to boost vaccination numbers turned out to be a precursor of what was to come. I was well aware the flu vaccine wasn't perfect. This is an elusive virus that continues to adapt to efforts to contain it. The vaccine probably was somewhere between 40 and 60 percent effective in preventing infection. So I was not surprised when people decided, "Why bother? I don't need the vaccine because it doesn't stop me from getting the flu." In fact, less than 50 percent of Americans get the vaccine annually.

But that's the wrong question. And it's the same wrong question that would dominate the next few years. The question that should have been asked was, "Will the flu vaccine effectively prevent me from dying?" The answer is that the vaccine is really good at that.

This was the message I wanted to get out: The flu vaccine is not only to prevent the flu, but also to prevent you from dying from the flu. The problem, I suspected, was the messaging.

I set as a goal finding out why people were hesitating to get proven vaccinations. This was still years before it was to become an issue during Covid. As I learned, there were several reasons other than they don't believe it works. The obvious one was that they didn't know about it. Then there were people who were misinformed, who had heard rumors that some vaccines actually cause disease. They were, understandably, afraid of making a mistake for themselves or the family. For other people, it was a problem of logistics. They couldn't take time off from work to get the vaccines they wanted. And a very small minority did not get vaccinated because they have a passionate aversion to vaccines.

This was still two years before the pandemic and the tragic, national debate about vaccines.

Equally important to me was preparing the country for the pandemic I was certain was coming, although I believed it would be a bird flu. Joy and I always felt that was perhaps the most important reason we decided to go to Atlanta. The position provided me an important, purposeful opportunity to use the diverse skills I had gathered in my career in a very focused way to prepare the agency to respond effectively to that challenge—whenever it happened.

The agency knew what to do, and we had plans on paper. Within a few months of becoming the director, I began practicing our response. I brought hundreds of people to Atlanta from around the country and the world. I created scenarios and taught people how to respond. In one of those exercises, for example, we claimed that a new potentially deadly flu virus was spreading rapidly. We put the participants in positions of power. How would they respond? What steps would they take, and in what order, to prevent infection? We did this to try to find out if there were holes in our program.

An amazing thing happened at that table. Within fifteen minutes it became real. We forgot it was an exercise. When we discussed numbers, big numbers, chills ran down my back. I remember I was forced to make recommendations about whether or not we should let elderly Americans get the vaccine because we didn't have sufficient needles and syringes to vaccinate the country against the new and more virulent flu virus.

We learned a lot by working through that scenario. We discovered we had a real problem about supplies. Where were we buying our needles? China. Where were we buying our syringes? China. And in this simulation, China did not want to sell us needles and syringes because they were fighting the same pandemic and needed them for their own people.

We discovered the holes in our program. Gaping holes. The problem was that we did not have the resources—or the necessary power—to plug them. The government doesn't like to spend money on speculation. There still are 1960s fallout shelters all over the country.

We also focused on bioterrorism. This remains a real and dangerous threat. In our scenario, a terrorist group built fake vending trucks and drove them through neighborhoods in several major cities spraying anthrax through its exhaust system. We looked at all the issues that came with it. How was the pathogen modified? Would the antibodies we had in our stockpile be effective? Did we have adequate treatment options?

In May 2018, the John Hopkins Center for Health Security held a simulation in which a genetically engineered weaponized novel virus they named Clade X, "for which there are no effective countermeasures," appeared in Germany and Venezuela and quickly spread around the world. I was invited to give the opening remarks. In this scenario, I explained, this disease, which "spread from person to person primarily by coughing and causes severe symptoms requiring hospitalization and intensive care in about half the people infected, overwhelms medical facilities."

Among the objectives of this exercise was to determine how the government "had to shift from making decisions concerning foreign diplomacy such as travel, monitoring and quarantining of those exposed, to domestic policies as they attempt to determine how to contain the virus and how to navigate the complex public health relationship between the government and private hospitals."

In this scenario, 150 million people around the world died, including 15 million Americans.

Many of us thought it was a very interesting exercise and took some level of comfort in the realization it was total fiction. It was possible, but still sometime way in the future. But we paid attention, we learned from it. Among the policy recommendations that emerged was the need to cut down the length of time it took to develop new vaccines and drugs for novel pathogens—they estimated it would take more than a year to produce a vaccine—as well as a need to strengthen national and global health systems prepared to respond to a pandemic. The scenario also outlined the steps the

CDC should take: develop a PCR test, issue travel alerts (in this scenario, for Germany and Venezuela, even though it would interrupt billions of dollars in trade), screen travelers returning from those countries at airports, issue alerts to health departments and hospitals, and isolate suspect cases and report them to the CDC.

We paid attention to these suggestions. We began taking steps to prepare for any type of health catastrophe. All of our response scenarios followed the proven path: detect, respond, and isolate.

Detect. It's impossible to overemphasize the importance of early detection. Before we could respond to Clade X, we had to know what we were facing. Detect. For the CDC, that meant expanding the agency's footprint on the ground. Stationing people in potential hot spots trained to look for anything unusual. We also worked closely with the intelligence community. For example, the government of Tanzania insisted there was no Ebola in that country. I had reports that was not accurate. Our intelligence agencies informed me that there had been an outbreak and Tanzania's defense department had managed to get it under control by quarantine and isolation.

The America government arguably has the best intelligence apparatus in the world. When I looked at global health security, we had some ability to identify potential threats, respond to them where they occurred, and shut them down so they didn't come into our space. That's the way we stopped potential Ebola outbreaks, for example. But the one exception to that, even if we did everything well, was a respiratory pathogen. A disease spread through airborne particles.

That's why early detection was so essential. I wanted to make certain we had access to the information that would alert us to potential problems and allow us to orchestrate a response. For example, if an intelligence agency suspected that a hostile country was bioweaponizing certain pathogens it would be our job to prepare for that.

Before becoming CDC director, I didn't realize how closely I would work with our intelligence agencies and how much highly classified information I would have access to. It turned out the government was well aware of the threat. I was constantly being brought into the highest-level security clearance discussions. As a physician, I was used to keeping patient information confidential, but this was on a much higher level. These were the types of issues that were so potentially catastrophic. It was almost impossible to sleep at night. We were considerably closer to dangerous situations with certain countries than most people could imagine. The fact that I could not share my concerns with Joy, which I always had done, made it even more difficult.

The National Security Agency had a pandemic team with a strong knowledge base of infectious diseases. Among them were a very smart virologist and a well-trained infectious disease physician. We tried to create channels that cut through existing red tape to make it easier to rapidly communicate information. But in May 2018, National Security Advisor John Bolton disbanded that team as part of an NSC reorganization. There was no real explanation. One anonymous administration official told reporters, "In a world of limited resources, you have to pick and choose." While several of those experts were still at the NSA, it was more difficult to access them.

Unfortunately, my predecessor was hamstrung by budget problems. I think it is fair to conclude that there was a general deemphasis on the importance of intelligence as part of our overall global security efforts. Several CDC biosecurity programs had been decreased or even shut down. Among those countries in which staffing had been reduced was China. Tom Friedan, who had served as director of the CDC during both Obama administrations, had used supplementary funding from Congress left over from the 2014–2015 Ebola outbreak to open biosecurity stations in seventeen countries. He gambled that Congress would grant that money. Unfortunately, when Fitzgerald became director that money was gone and Congress would not support that program. As a result, many of those programs were shut down, while others were decreased. China was among those places where the number of Americans on the ground was decreased.

That made absolutely no sense to me. If a new dangerous pathogen was going to emerge anywhere in the world, China was right at the top of my list. They had all the elements necessary to create a new pathogen: a vast agricultural economy, rudimentary hygiene in certain areas, and sophisticated military and civilian biological research programs.

I reversed that decision. I believed we needed to augment, replace, and increase our presence in China. That's where we needed more eyes and ears. I let the agency know that was a priority. But it was still subject to the budget we had, which was not great.

This was still two years before the pandemic began and questions arose about why we weren't better prepared for COVID-19.

The problem I encountered was how to pay for that early warning system. On spreadsheets, the CDC was well funded. Congress understood the importance of our role and gave us billions of dollars. When I took control, I asked the chief financial officer if they had outlined the director's budget. In my various supervisory roles at the university and the Institute, I was familiar with working within a budget, I just needed to know what it was.

"You don't have a budget," she told me.

I was shocked. "What do you mean? I'm the director of the CDC. I must have a budget."

"No," she said, "you don't."

As I was to learn, there was no overall CDC budget. Instead, Congress funded specific programs. There probably were three hundred different funding lines. There was money to be used for maternal survival programs. Money to study pig diseases. Money to study vaccines. But there were no director's discretionary funds.

There were different ways I could get money when it was needed, but it was a process. I wanted a funding line available to the director to invest wherever it was needed to further the core capabilities of public health both at home and internationally. That did not exist.

One of my first efforts as director was to try to change that. As I eventually testified to Congress, "it is important to realize that for decades we have underinvested in the public health infrastructure of this nation. . . . CDC's funding that we get from Congress, about 70 percent goes out to local and state, territorial, and tribal health departments."

Incredibly, the CDC director had no direct access to money needed to respond to emergencies. If we had to respond in real time to a dangerous situation, say a strange virus coming out of China, there were no provisions for that. Instead, we had to go to the CDC Foundation, an independent organization, which then raised the needed money from private-sector donations.

The CDC Foundation had been created by Congress in 1992 to raise funding for the agency. I didn't know for certain, but I wasn't aware of any government agency so reliant on an independent group for money. When infectious diseases are spreading, time matters—and money can buy time. We certainly were delayed mounting a coordinated response to the first Ebola outbreak because we didn't have the money we needed.

I worked with Congressman Tom Cole and Senator Roy Blunt, both of whom served on appropriations committees, to try to make rational changes. The existing system is nonsensical, I told them. When we are trying desperately to stop the spread of an infectious disease, we don't have time to be begging for donations.

They led the fight in Congress and eventually about $40 million was made available in a rapid response fund. I used it for the first time during the Ebola outbreak in the war zone. Over time, the fund increased to almost $350 million. That was very fortunate. When the COVID-19 pandemic began, we were able to immediately start designing a response. It wasn't perfect, but I believe our actions made a difference in the horrendous body count. That also gave me the funding we needed to attack other growing problems.

Many people believe the only function of the CDC is to protect the country against outbreaks of disease, which was the primary mission when Malaria Control became the Communicable Disease Center. But it had expanded well beyond that. Before settling into my office, I'd made a list of the things I wanted to accomplish my first year. I had so many great plans. In addition to preparing for unexpected challenges we might face, I wanted to go on the offensive. I wanted to attack existing health issues.

The purview of the director had expanded to cover just about every aspect that contributes to American health outcomes. To get the things on my list done, I needed the support of Alex Azar, the secretary of health and human services. I knew how important it was to build a strong working relationship with him. And if we did have to face a major challenge—a pandemic, for example—it would be vital that we were able to work together.

Unfortunately, that relationship was rocky from the beginning. Within days of starting the job, my salary became an issue. The media reported that I was making almost double what my predecessor had earned, far more than Azar, other cabinet secretaries, and even the head of the NIH.

I had no idea any of that was true until I read it in the newspapers. Honestly, I had never negotiated any salary. I didn't know what I was being paid (which, by the way, infuriated Joy). I only found out I was to be paid $375,000 when I began filling out financial forms. Problems started when Senator Murray sent a public letter to Secretary Azar wondering "why someone with limited public health experience, particularly in a leadership role, is being disproportionately compensated for his work."

I hadn't taken the job for the money. In fact, it was slightly less than half of what I had earned the previous year. But about a week after I had been sworn in, the secretary's chief of staff called me into his office on Friday afternoon and asked me why I had privately negotiated my salary with the White House.

"What are you talking about?" I responded. I told him I had not known what the salary was when I accepted the job, adding, "my wife was pretty upset with me that I didn't have that locked down." My salary had been set and approved under a provision that allows the government to pay employees salaries competitive with the private sector. The situation was very upsetting; he felt I had somehow squeezed an exorbitant salary out of the government. In fact, in that meeting I was told Secretary Azar made it clear that if he had known he needed to pay me that much money to take the job he probably would have looked for somebody else.

That whole impression, which reflected poorly on my reputation, really bothered me. Could anyone actually believe that after spending my entire adult life practicing medicine, after throwing myself into the middle of the AIDS plague, that suddenly I was somehow in it for the money? It was an insult. It wasn't true in any way. Two days later, I called Azar and calmly told him, "Mr. Secretary. I don't want my salary to get in the way. I'm disappointed this has become an issue. I don't want to die by Chinese water torture."

"I also don't want this to interfere with our relationship. I want you to go ahead and change my salary to whatever you think is appropriate. It's about the mission, not the money." My salary was reduced to $165,000. But I was surprised a few weeks later when I found out who had initially approved my salary: Secretary Azar. He actually had signed the approval.

So I should have known this was not going to be the easiest path. Professionally, it actually started out very well. Usually, when I made suggestions supported by data, he could be very supportive.

Among my goals, for example, was to finally end the AIDS epidemic in America. If I could get that done, I believed, no matter what else happened my term in office would have been a success. Unless, of course, there was a pandemic.

When I first presented my plan to him, he described it as "aspirational," rather than realistic. But when I presented evidence to him showing that 50 percent of all infections occurred in only forty-eight counties, Washington, DC, and San Juan, Puerto Rico, he suddenly agreed that this could be manageable and became a great help getting the president to initiate that initiative.

Azar also strongly supported my second initiative. Data had landed on my desk showing that 37 percent of high school students and about 20 percent of middle school students were using nicotine products on a daily basis. That was extremely disturbing to me because I have eleven grandchildren.

Nicotine is a poison. It's an addictive drug. It negatively impacts brain development. The surgeon general's original attack on smoking cigarettes had been successful. The number of young people addicted to cigarettes had decreased. In response, the tobacco companies had introduced vaping. The use of a Juul, an electronic device resembling a metal cigarette, which mixed nicotine and a flavoring to create a pleasurable mist, became widespread. While the Juul was originally sold as a tool that might help cigarette smokers break their addiction to nicotine, it did precisely the opposite. Moreover, manufacturers specifically manufactured sweet and fruity flavors, like cotton candy and Juicy Fruit, to attract young users.

What was not generally known was that the amount of nicotine delivered in one inhaled Juulpod was as much as ten times that of a single cigarette. The result was that kids were getting addicted to nicotine within days rather than months. I went to Azar and told him we really needed to stop this, we needed to push back.

He agreed. I thought we were building a good relationship. Along with the head of the Federal Drug Administration, we met with the president in the Oval Office, urging him to make a strong statement by taking these flavored nicotine products off the market.

Initially, he agreed. That was very exciting. It would be a major achievement. But gradually the administration backtracked. For example, menthol, which made these products more desirable in certain populations, including African Americans, was excluded from the directive. Eventually the entire initiative was watered down to have limited impact.

My relationship with Azar never developed the way I had hoped. The biggest problem probably was the most important: communication. His management style was designed to deflect responsibility. He did not want me to text him, email him, or leave a message. He was a lawyer; he did not want to create a paper trail. As far as I was concerned, and this is only my opinion, he went out of his way to avoid responsibility. When I wanted to speak with him, I was supposed to call him, let the phone ring several times, and then hang up. He would call me back.

When we had something important to discuss, we would do it in person. I can't remember a single meeting with other agency heads present during which we discussed anything substantive. I understand many people have issues with their boss. And, like most people, I'm not interested in hearing about it. You have a job to do, do it. But I am describing our relationship because I strongly believe it negatively impacted our response to the pandemic. The lack of direct communication cost us time—and that was a luxury we didn't have. It's an important lesson. Equally damaging, he also took control of all CDC communications. For example, he tried to edit scientific publications, and then he became the self-appointed, on-air spokesperson for the agency. In the early days of the pandemic, Margaret Brennan invited me to appear on the Sunday-morning show *Face the Nation*. It was an important opportunity to speak directly to the American public. I wanted to explain what we knew and what we were doing. I wanted to offer the best advice we had at the time. I wanted to calm the rising tide of panic. As I was getting ready to do the show, I received a call from Azar's office telling me to stand down. The message was clear: CDC directors don't do *Face the Nation*. Secretary Azar will do the interview.

Our relationship started poorly and gradually deteriorated.

This began almost two years before the pandemic.

Believe me, we were not caught by surprise. I had spent a considerable amount of time during that first year preparing to respond to a pandemic. I knew that the biggest challenge we could face was a respiratory pathogen that had significant mortality and morbidity. Literally two weeks after I went to work, we scheduled a pandemic flu exercise, which would "focus on the emergency response process at CDC to include discussion of past responses, including H1N1 (bird flu) and future potential responses."

A month later, only days before the Johns Hopkins Clade X exercise, the CDC and Emory University hosted a symposium to commemorate the centenary of the Spanish flu. Among the panel discussions and presentations on the agenda were "Current Gaps; United States and Local Public Health Readiness" as well as the question I was continually asking, "Are We Ready to Respond to the Next Pandemic; United States Government Preparedness Posture in 2018."

I did everything possible to spread the warning. Sometimes it felt like I was shouting into a hurricane. In October, I spoke at the Milken Future of Health Summit in New York. Asked about my priorities, I told the audience that under my direction the CDC was going to reverse the opioid epidemic—what I called "the public health crisis of our time." We were going to cross the finish line to eliminate polio. Moreover, we were going to ramp up vaccinations and focus on responding to disease outbreaks, ensuring global health security. "I'm deeply concerned about pandemic and seasonal flu," I explained, and "flu remains our number one threat."

That had become my mantra. At the end of October, I appeared on *CBS Morning News* to urge people to get the seasonal flu vaccine. In response to a question from John Dickerson, I said, "People ask me what keeps me up at night, and the answer is what you just brought up, pandemic flu. I think it's very possible we're at risk for another pandemic. Our best preparation for that pandemic is to optimize response to the seasonal flu. . . . The most powerful tool we have to eliminate disease is vaccination."

So when the phone rang on December 31, 2019, I was concerned. But I was neither surprised nor alarmed. Not yet. This was step one in action: detection. This was the Paul Revere of American public health sounding a warning. We had to act quickly to find out exactly what this new challenge was. The proverbial million thoughts raced through my mind in the next few minutes. I had been involved in situations like this several times as an expert in infectious diseases, but never in the decision-making position. I asked the obvious questions: Exactly what did we know about this outbreak? What access did we have to these patients? Did we know the nature and extent of their symptoms? What was the Chinese government saying about this or were they ignoring it? How sure were we that this was limited to people who had been in that market? The information was sketchy, at best.

I followed the procedures we had put in place. All we knew was that we had become aware of twenty-seven cases of an unexplained pulmonary illness among people who had been in a wet market in Wuhan, China. The World Health Organization apparently was already aware of this. Within several hours, I had gotten the written reports so I could review each case. That's when I became aware that there with three instances of multiple infections within a family unit. Three clusters. That's a strong indication that whatever this thing was, it was acting like an infectious disease, spreading from person to person. That was the first red flag.

Either that day or early the next morning, I called my counterpart in China, George Gao, director of the Chinese Center for Disease Control and Prevention. It is a very professional organization. Some very good people work there. Fortunately, I had known George for years. We spoke the same language: pathogen. He had been trained in the West, including several years at Oxford and Harvard. Having an existing relationship with him cut through a lot of time-consuming issues; at that point, there was very little politics involved. Viruses are not stopped by borders or now, with air travel, by great oceans. Whatever this was, China would be dealing with it too. In fact, they already were.

We spoke several times during the next few days. George had only found out that there was a respiratory outbreak of an unspecified pathogen at most a few days before me. He legitimately did not know, as we would later learn, that in September or October there had been a huge outbreak of respiratory illness in a lab in Wuhan. Or that it had begun spreading. It was not until December 31, the same day I learned about this, that Wuhan government officials posted the very first notice that there was a medical situation, admitting they were "investigating a pneumonia outbreak."

George Gao and I were both hopeful that it had been restricted to the twenty-seven people in the wet market.

During our discussions, I offered to send him twenty to thirty people from the CDC, plus my people already in Beijing, to try to get a handle on this. First and foremost, I asked him, is there human-to-human transmission? Yes or no. Second, if it was human-to-human, was that transmission dependent on severe illness symptoms or was this virus transmissible without symptoms?

George assured me on the phone that he had seen no evidence of human-to-human transmission. He was confident that this was spread from an animal to a human. That was possible; several people might live in a home or on a farm caring for the same animal. It was possible, but unlikely. "George," I told him, "I'm very concerned about that conclusion." We talked about it for a while, and he agreed to go back and take another look.

The next time we spoke, he told me that several infected people had been hospitalized. I don't remember the number, only that he had investigated and could confirm there were no nosocomial transmissions. This meant none of the healthcare providers had been infected.

"I'm still very worried when I see these clusters," I said. "Look, right now we're only looking at people infected with this thing who have had some exposure to the wet market. Every single one of them has been there. Let's see if that's the extent of it. What I'd like you to do is go look for people who got this thing independent of the wet market. That'll tell us a lot."

George agreed. Two days later, he called me back. "Basically," he said, "we're in trouble. We have a very serious problem. This is much broader than I thought." Then, as the enormity of what we were now dealing with hit him, he broke down and became emotional. He knew Wuhan was dealing with a major respiratory epidemic.

The first stage was done: A new respiratory disease had been detected. My immediate response was that we needed to get some of our people on the ground in Wuhan, people who could quickly answer some critical questions. George Gao asked me to write a letter officially offering to help, and he would try to get an invitation from the Minister of Health. I briefed Secretary Azar, who was highly supportive. He reached out to China's Minister of Health for assistance.

We waited. I contacted George Gao every day about the invitation, and every day he told me he had not gotten approval. Meanwhile, this thing, we still didn't know what it was, was spreading.

On January 2, I gave the first briefing to the National Security Council. They got it right away. They understood what we potentially were facing. The counterproliferation and biodefense directorate, the so-called Weapons of Mass Destruction unit, took charge. They created models that predicted mass casualties if we did not respond aggressively. I met with this team every day as we tried to track its spread. For several weeks, it was basically me and the NSC watching this thing. We still had no known cases in America.

On January 11, China officially announced the first death from this disease.

At the time, we did not realize how unprepared we were.

It turned out that all that time I had spent that first year trying to spread the warning had little impact. While the CDC led the response, we did not have the workforce that we needed. We did not have sufficient quarantine and isolation facilities. We did not have the budget to pay for any of it. We did not have the relationship with the diagnostic private sector that we needed. We did not have an aggressive development program ready to immediately begin developing therapeutic countermeasures. We did not have a clear chain of government command; indeed, it was hard to know who the decision maker was for specific needs. We certainly did not have an adequate communications system capable of spreading either the alarm or the needed information to the public.

In short, America was woefully unprepared for one of the greatest pandemics in the history of the world.

FOUR

There is a scene in many monster films in which a technician rushes into a laboratory and exclaims with shock and fear, "It's reached the edge of the city!"

This disease had reached the edge of the city. And it was spreading. On February 10, WHO announced that this virus had killed 1,013 people worldwide—still mostly in China—which was more than the SARS outbreak that had terrified the world in 2002 and 2003. A day later, that organization gave the enemy a name, COVID-19. It was an abbreviated description of Coronavirus 2019.

We still were operating with the belief that we could contain this virus by identifying symptomatic cases, isolating those cases, doing the necessary contact tracing, and then isolating those contacts. That had worked before. I remember after we had the first case in Seattle I spoke with Washington's secretary of health, Dr. John Weisman. We had become close while dealing with the AIDS epidemic and curtailing drug usage. John explained that his department had stopped conducting contact tracing. "What do you mean?" I asked. "We've got to contain this pathogen."

"Bob," he told me, "I just don't have capacity."

He told me he needed twenty or thirty more people. "I'm sending them to you tomorrow morning," I said. "We need to contain this pathogen."

Several days later, I got the exact same call from the California Department of Public Health. They did not have sufficient staff to conduct contact tracing. I immediately sent some of our people there. The response of the CDC personnel was heroic. These were lab technicians, epidemiologists, and data collectors—all of them volunteers. These people literally put their lives aside with only a few hours' notice to fly into the middle of a spreading disease to protect the American public.

But about a week later, I got another call from the California Department of Public Health. This one shook me. They were reporting that a nurse was infected. She had not been in contact with anyone known to be infected. A day later, a second nurse in the same community tested positive. That was the day I realized we had community transmission in the United States. This virus was no longer confined to China or even Asia. We were seeing active transmission in the United States.

I began searching the databases we normally monitor. I saw that influenza-like cases in New York that had been going down, down, down suddenly were going up, up, up. But the number of diagnosed cases of flu were still going down. I called Harold Zucker, commissioner of the New York State Department of Health. I told him, "Harold, we've got a problem in New York City. I think you've got Covid there." This was before the first case was confirmed. Within a week, it was spreading there.

Among the CDC's varied responsibilities is maintaining quarantine stations at twenty ports and land crossings where international travelers enter the country. Basically, our people there assess tourists as well as returning citizens and animals for health issues. When necessary, they isolate and even quarantine them. If they didn't have symptoms or a high temperature, they were out the door. As the number of cases began multiplying, we began opening additional isolation centers on military bases. In February, we screened about 50,000 people. A woman who had returned from Wuhan was quarantined in San Antonio, Texas, for two weeks, but after two consecutive tests taken at least a day apart came back negative she was released. She headed for the airport to fly home. After she had left though, her third and final test was "weakly positive." She was still infectious. And we couldn't find her.

We believed we had a potential carrier about to get on a crowded airplane. I spent much of the night and early morning on the phone as authorities there worked desperately to track her down. They finally located her in a hotel and put her back in quarantine. Our people discovered she had contact with about a dozen people in a hotel. We began monitoring those people, but none of them were infected.

Our strategy, my strategy, was correct for what we knew and when we knew it. For the first two months or so, I believed we had this under control. I was out there in public urging Americans to block and tackle, just block and tackle. Identify symptomatic patients, isolate their contacts, and shut it down. If we could do that as we had done in the past—boom, boom, boom—we'd be on the right track. As I told a congressional committee in late February, while "the immediate risk of this new virus to the American public is low, CDC is preparing the nation's healthcare system to respond to identification of individual cases and potential person-to-person transmission of COVID-19 in the community."

Meanwhile, we were completely oblivious about what really was going on. We were missing 50 to 70 percent of all infections. And then we learned about the *Diamond Princess*.

It's not a sin to be wrong in science. You begin with a hypothesis. You promote it, debate it, investigate it and test it. You compile data and examine it. It's a process. The data will show that the hypothesis is right or wrong. If it's wrong, you acknowledge it, and then you test it again, and again, until you get it right. That's science. Throughout my career, I've followed the data. This situation was no different. The original characteristics of this virus led us to believe it was a version of SARS and MERS. The data made it clear we were looking at something entirely different.

The *Diamond Princess* was a British cruise ship docked in Yokohama. It was carrying 3,700 passengers and crew members—among them about five hundred American citizens or permanent residents—on a two-week cruise to China, Vietnam, and Taiwan. On February 2, the captain had been informed an elderly passenger who had left the ship a week earlier in Hong Kong had been diagnosed with the same viral disease that had caused China to close Wuhan. A day later, health workers discovered that nine passengers and one crew member were infected. To prevent the disease from spreading in Japan, which had only about twenty known cases, the ship was quarantined for at least two weeks. It was the worst situation imaginable: an unknown number of infected people forced into an enclosed area. As we advised, passengers were confined to their staterooms as much as possible, but the crew had to continue circulating throughout the ship serving food, distributing clean linens, and taking care of other necessities. The corporate medical officer emailed the crew: "What is happening is unprecedented, but it is allowing health experts to learn about the virus and how it spreads. This will help all of you onboard, as well as people around world." Basically, they were admitting, the *Diamond Princess* was essentially a floating petri dish.

Within days, the *Diamond Princess* had become the largest outbreak of the disease outside China. We sent teams to Japan to monitor the situation. Day after day, an increasing number of people were infected. After almost two weeks trapped onboard, more than three hundred Americans who tested negative were flown home, where they were quarantined for an additional two weeks. In Japan, many infected people were transported to local hospitals.

Initially, we only tested people showing symptoms, but by February 11, we had started testing everyone. Conditions aboard the ship were deteriorating, and we intended to continue isolating infected passengers while allowing healthy people to be released after a brief isolation. Within a few days, I was looking at test data that stunned me: On February 20, tests confirmed we had 619 cases on the ship, and almost half of them were asymptomatic.

I knew instantly what that meant: We were dealing with a new and different virus. A virus that could be communicated by asymptomatic people. In fact, a significant number of those asymptomatic people aboard the *Diamond Princess* never developed any symptoms. For most of human existence we did not know viruses could infect people asymptomatically. We didn't know how to look for them, or even if we should. But the tools of modern science—antibody surveys that can detect past infections, tests that find viral DNA or RNA in apparently healthy people, and mathematical models that tell us about viral progression—made the previously invisible visible.

Looking at those numbers only emphasized the frustration I felt that we weren't in China the first two weeks of January. That had made all the difference. If we had been in Wuhan, if we had access to patients, we would have figured it out. Was there human-to-human? Was it asymptomatic? We would have been able to set up a strategy based on the reality that asymptomatic transmission was going to be the driver of this disease. It took me another three weeks to figure that out. The fact that there had been a limited number of cases in America through February made us believe we were following the right strategy. I had about eight hundred contacts that we evaluated, and only two of those people ended up testing positive. So I concluded, mistakenly, along with the CDC, that this wasn't so infectious, despite what we were hearing from China. Our mistake was that we evaluated them by asking them if they felt sick, and only if they had symptoms did we test them. If they weren't symptomatic, we didn't test them.

Had I known differently, I would have been asking for a Manhattan Project—a complete government effort—to maximize testing capability. That was a vitally important lesson: If we had relied on test results rather than symptoms in February and early March, we would have been way ahead in identifying people who were transmitting it.

The *Diamond Princess* changed everything.

It was obvious that we could not prevent COVID-19 from spreading. In mid-February, I met in the CDC conference room with five or six of our people to assess the potential impact. We still had fewer than fourteen cases and no deaths. Our modelers, our statisticians, sat me down and told me that their model predicted that by the following September this pandemic possibly could kill 2.2 million Americans.

For several seconds, the room was absolutely silent as that number sank in. 2.2 million American deaths? This was going to be catastrophic.

These people were not trying to frighten anyone. They weren't exaggerating. These were experts, very objective individuals. This is what history and statistics were telling us. I thought that number was very believable. I knew the potential of respiratory pandemics to cause a catastrophic number of deaths. I remember photographs I had seen of Philadelphia and other cities during the 1917 pandemic. One that stuck in my mind was a fire station with bodies piled in front of it.

At home that night, I confided in Joy. There were tears. I knew that people in their sixties and older would be disproportionately affected by this because in many instances our immune systems are already compromised. Joy and I accepted the fact that there was a real chance we both would die before September. We weren't frightened by that; but without doubt, we leaned heavily on our faith.

I informed all the members of the task force, which included the president and vice president. I stressed that these numbers were not exaggerations. This was what we were facing I remained hopeful that we still might have been able to contain it. I told myself it was still possible to keep this at ten thousand cases. That was my objective: ten thousand cases.

The line between informing the public and scaring them in a situation like this is a thin one. Because the disease had not yet spread widely, I wasn't confident Americans fully grasped what we were facing. Admittedly, at that moment we were medically helpless, there were no viable countermeasures. We had no vaccines or antivirals to try. Our only real defense was to try to get the American public to be more careful about exposure. So if we were going to contain this thing, we needed their full cooperation. In late February, while there still were only fourteen cases in the country, we began preparing the country for what we now knew was coming. "COVID-19 is a serious public health threat," we warned in our daily update. In that release, we noted that the Chinese had already taken the first draconian steps to mitigate the spread, closing major transportation hubs, preventing people from leaving cities where Covid was spreading, closing down schools, closing down factories, and even canceling New Year's celebrations.

As calmly as possible, we suggested similar measures that might be imposed here, including "school dismissals and social distancing," "postponement or cancellation of mass gatherings," and "telework or remote-meeting options in workplaces." We recognized "These measures can be disruptive and might have societal and economic impact. . . . However, studies have shown that early layered implementations of these interventions can reduce the community spread and impact of infectious pathogens . . . even when specific pharmaceutical treatments and vaccines are not available."

In a telebriefing for the media, the director of the CDC's National Center for Immunization and Respiratory Diseases, Dr. Nancy Messonnier, began spreading the message. "Ultimately, we expect we will see community spread in this country," she announced. It no longer was a question of "if" we were going to be impacted by this virus, just when and how severely. COVID-19 had already met two of the three criteria needed to be classified as a pandemic, she explained. It was potentially lethal and was spread from person to person. The third factor was worldwide spread.

Already, it had spread to thirty countries and territories in addition to the United States and China.

We knew how cultures throughout history had fought infectious diseases, we understood what had worked or failed, but those lessons had little value to us. Never before had a world, in which masses of potentially infected people traveled so easily and quickly between nations, had to deal with an easily communicated infectious disease. Never before had the world economy been so dependent on rapid transportation, which was going to be severely impacted.

Without vaccines or antivirals, we had few options. The only thing we could do was try to curtail its spread by identifying and isolating. The meant we needed more tests than we had ever anticipated, tens of millions more, far more tests than our labs were capable of producing. Any belief we had that this could be managed by the public health sector was gone. We needed the full involvement of private companies. Several smaller companies had been producing commercial tests, for which they were charging between $18 and $100 for a two-pack. The World Health Organization had begun distributing 250,000 kits produced by a German firm to laboratories around the world. But widespread production and distribution of tests here was hampered by FDA quality-control restrictions. Even after public health labs figured out how to remove the third set of primers in our test to make it effective, the FDA refused to approve it.

That all changed on February 28. We invited representatives of Big Pharma to a meeting in the Roosevelt Room of the White House and laid out the situation. The FDA announced it was relaxing its guidelines, allowing labs to use our test. Within a week, Congress had passed its first Covid appropriations bill, which provided $8.3 billion in emergency funding for public health agencies and coronavirus vaccine research. Of those funds, $7.8 billion went to federal, state, and local health agencies. Big Pharma got the message: There was a huge amount of money available to fight this disease.

On March 12, Roche announced the FDA had granted it emergency use authorization for a swab sample test that could deliver results in three and a half hours, much faster than existing tests.

While that was going on, we began shutting down the world. That was the only weapon we had. As an example of what was about to happen, at that moment very few people outside the business community had even heard of a relatively obscure free communications website named Zoom. The initial price offering of this "videoconferencing tool" in April 2019 was $36. By the end of 2020 the stock would be selling for $383 a share.

We were learning more about this virus every day. The *Diamond Princess* taught us that a cruise ship was an incubator. Fourteen people aboard that ship had died. We were getting reports of outbreaks on other ships carrying Americans. On March 4, we learned that an American who had been aboard the *Grand Princess* two weeks earlier had died. We began testing. Vice President Pence announced that of the forty-six people tested onboard, twenty-one of them were positive. It turned out that about 80 percent of all the cruise ships at sea had outbreaks.

That was logical. A cruise ship is the perfect environment for an infectious disease to prosper. Thousands of people, many of them over sixty-five, are sailing together for a long time in a relatively closed environment. They share cabins. They gather in the ship's cafeterias, restaurants, health club, and theaters. Then they disembark, thousands of them, bringing home with them to family, friends, and neighbors whatever they may have caught. The crew was even more susceptible, having daily contact with numerous passengers and living and eating in small, confined spaces. And one of the most surprising things we discovered was that the virus could be spread horizontally from the top deck to the lowest deck, cabin by cabin, by the connected sewage system. Toilets were designed so that a flap went up when it was flushed, which created an aerosolization, a mist that spread Covid.

This was my first enemy of the people moment. In 2019, the cruise industry contributed $55.9 billion to the American economy. The 13.7 million passengers embarking from American ports created almost 500,000 jobs and generated $25 billion in wages. It wasn't only the crews of these ships who would be affected, tens of thousands of other people provided the food, handled the baggage on the docks, and shuttled passengers back and forth from airports. Their income depended on this industry.

And I had decided I had to shut it all down.

The economic consequences were enormous. Before actually issuing the order, Vice President Pence and I flew to Fort Lauderdale, where we met with the CEOs of all the major cruise lines; they fought back, trying to convince me they could find a way to continue to sail safely. I just didn't believe that was possible. The environment made that impossible. I also met with many Florida politicians, including both senators and several members of Congress, who told me this would devastate that state's economy. I spent hours trying to find some way to keep this industry operating that made sense. There was none. As CDC director, this was my decision. My job was to protect the public health.

There were a lot of people who believed this was excessive, that the limited number of infections and the small number of deaths, forty-one at that time, just didn't justify this decision. I understood that, I got it. I just didn't agree with them. It was a tremendous decision. It was one I had never asked for or wanted to have. To save lives, I would have to put tens of thousands of people out of work.

On March 11, the World Health Organization officially declared COVID-19 a pandemic, meaning the entire world was at risk.

The president spoke to the nation that night to explain it, telling Americans, "This is the most aggressive and comprehensive effort to confront a foreign virus in modern history. I am confident that by counting and continuing to take these tough measures, we will significantly reduce the threat to our citizens, and we will ultimately and expeditiously defeat this virus." He then announced he was imposing a partial travel ban, prohibiting non-Americans who had been in Europe within the previous two weeks from entering the country. That was a very controversial action.

It was as if a dark cloud was slowly covering the sun. That same night an NBA game in Oklahoma City was suddenly postponed when a player tested positive. It was soon announced that the league's entire season was being suspended until further notice. Mark Cuban, the owner of the Dallas Mavericks, told reporters, "This is much bigger than the NBA. Do we send our kids to school tomorrow? Is it that big? It's like out of a movie. It doesn't seem real."

On March 14, I issued the No Sail Order. "Based on substantial epidemiological evidence related to congregate settings and mass gatherings, this order suspends operation of vessels with the capacity to carry two hundred fifty individuals or more. Evidence shows that settings as small as nursing homes or movie theaters can proliferate the spread of a communicable disease." I outlined the dangers to the general public, "Cruise ship travel markedly increases the risk and impact of the COVID-19 disease outbreak within the United States. Disembarkation of passengers at sequential ports may lead to disease transmission in those ports. Return of disembarked infected passengers to their communities could lead to widespread disease transmission." Therefore, I went on, "cruise ship operators shall not embark any new passengers or crew. . . . Cruise ship operators shall not commence or continue operations."

I made it clear that this was only a temporary ban, just until we could design ways to ensure the continued safety of the passengers, crews, and all the people they would encounter in the days following their return. I was willing to meet with industry representatives whenever they were ready to meet. I absolutely wanted to reopen as soon as possible, I just didn't see it happening very quickly.

I had been speaking regularly with Vice President Pence and other members of the administration. The projected 2.2. million deaths certainly made an impact on them. I was among a group of people urging them to follow the lead of other countries and take harsh steps to mitigate the spread of COVID-19. We believed, though we certainly did not know, that by clamping down China had gotten the epidemic under control. We were told they were reporting only fifty new cases the previous week. There was so much we didn't know. Initially, the CDC discouraged gatherings of more than two hundred fifty people. We didn't mandate that. We just suggested that people stay out of places like bars, restaurants, and food courts. I don't remember how we reached that number, but obviously it made little sense. We reduced that to fifty people.

There was considerable debate about what additional steps needed to be taken to cut down transmission. There was no good data to guide us. We had to rely on history and experience. The final decisions we made were based on policy rather than science. When numbers don't tell a story you have to ask logical questions. The most overriding question was what steps do we take that will save the most lives?

There was no right answer. Good, smart people argued all sides of the equation. It eventually was decided to allow each state governor to make that decision for their constituents.

These many years later, there still is no "right" answer to that question.

There was a general agreement that we should basically put the entire country on pause for fifteen or thirty days while we tried to figure out a sensible way of moving forward. I was not in favor of shutting down our economy and schools. I was pretty outspoken against it, especially shutting down schools. I didn't believe it would get us where we needed to go. There still was no evidence that schoolchildren were bringing home infections, and I was concerned the potential long-term damage to the educational process might have a long-lasting, profound influence on those children's lives. As for the economy, while clearly there were areas that needed to be greatly curtailed or shut down, I also believed there were other areas where the danger could be mitigated. I thought we needed to be much more surgical in our mitigation decisions. I argued pretty strongly that it was easy to shut something down, but it was hard to open it back up.

Tony Fauci and I were asked to testify in front of the House Oversight Committee. Ironically, the committee room was filled with people sitting side by side as we testified about restricting social contacts. Tony Fauci was becoming the media's go-to expert; he began to consider himself the public health voice of America. Then Deborah Brix was appointed head of the task force, and she considered herself the public health voice of America. Then Alex Azar, secretary of HHS, decided he should be the public health voice of America. Honestly, as director of the CDC, I also considered myself the public health voice of America.

That became a significant problem. When facing a crisis, the government absolutely needs to speak with one voice. We had different people stating different opinions—the lack of scientific knowledge limited us to our opinions—and that further confused the public. The media exacerbated the problem. Years earlier, a select few trusted people delivered the news to the American public, but those days of Walter Cronkite, David Brinkly, and Robert MacNeil and Jim Lehrer were long gone. That was long before cable TV channels began competing with the networks for viewers, and the internet made it possible for anyone to spread any claims, no matter how wild. Newspapers, which had been the primary source of information, were fighting for their existence. As a consequence, all of these media outlets were trying to attract a segment of the audience with sensationalism and reporting that appealed to their beliefs.

The accuracy of what was being reported was basically a flip of the coin. Or a change of the channel. Or a click to a different website. It was possible to find some level of support for whatever views a person already held. The creation of the virus was a government project to establish a dictatorship? That was there. COVID-19 was a Chinese bioweapon? That was there. The virus was created to attack only certain ethnic groups? That was there. Too often what was being reported, on both sides of the political spectrum, was totally inaccurate. Unfortunately, rather than getting accurate information from a politically neutral source, too many people relied on outlets aligned with their own political beliefs. As a result, unscientific points of view began spreading, too often playing on fear. When we reported information that contradicted what people were being told, we were accused of being biased or worse, dishonest. Too many people believed we had a political agenda, which absolutely was not true. The CDC does not have a lot of power to force compliance; we depend on persuasion. As a result of all this, when we most needed the public to follow sensible guidelines, to work with us to mitigate the spread, people hesitated. They didn't know what to believe.

Tony and I had been colleagues, if not close friends, for decades. He has twice nominated me for membership in the National Academy of Sciences, the nation's most prestigious scientific organization. A few years before this, he had invited me to give a major lecture at NIH. On several previous occasions, I had reached out to him for advice, and we'd had a fine scientific discussion. There was a great deal of mutual respect between Tony, Deborah Brix, and I. When we disagreed, we would discuss the issue, not dismiss the other person's opinion. We never doubted that our opinions were grounded in science and an understanding of the data. We discussed why we saw things differently. That generally resulted in high-level, usually effective scientific dialogue among knowledgeable people who respected each other. In fact, I think we really valued that, as it forced us to consider divergent points of view. So when we confronted this mystery virus, it was not a surprise that we reached somewhat different conclusions, or that we were unafraid to have a dialogue about our different conclusions.

But while it was to become obvious that Tony and I had some substantial disagreements on how to move forward to mitigate the consequences, we certainly did agree about what we were facing and the dangers it posed. At this congressional hearing, New York Representative Carolyn Maloney asked, "Is the worst yet to come?"

"Yes, it is," Tony responded. "I can say we will see more cases, and things will get worse than they are right now. How much worse will depend on our ability to do two things: to contain the influx of people who are infected coming from the outside and the ability to contain and mitigate within our own country."

"Bottom line, it's going to get worse."

Every day presented a bigger, more complex challenge. Joy and I were living in a small apartment in Atlanta, right across the street from the CDC. We had two bedrooms; we slept in one, and I used the second room as an office. That office overlooked the entire CDC campus, so I never really escaped that place. I would come home with fifty, sixty, one hundred pages of material to read through. I stayed up late into the night and often was up by 5:30 in the morning. I was always on the phone. Joy remembers a night I feel asleep in the middle of a 3 a.m. call with, I believe, the governor of Texas.

Usually around 8 a.m. I'd walk over to my office at the CDC. At 10 a.m. every morning, seven days a week, whether I was physically there or on the phone, I would meet with my in-house task force.

We would patiently go through much of what we had learned the day before, constantly trying to see if there was something else we needed to do, if there was a different approach we might consider, if there was anything we were missing. Should we be rethinking yesterday's decisions?

I was privileged to have an extraordinary staff. Without exception, these people were totally professional public servants, committed to public health. These were the kind of people you wanted to go into battle with. So when I heard the agency criticized, I could get really angry. Of course, mistakes were made; in science making mistakes and learning from them is an essential part of the process. But when politicians attacked us, at least some of the time for their own political benefit, I was not afraid to speak up for the people who were risking their lives to win the battle against COVID-19.

Americans were getting progressively more frightened and angry. We are not used to being forced to change our lives or, in many instances, sacrifice comfort for the benefit of others—especially when we aren't convinced the people in charge know what they are doing. This country has been so successful for so long meeting so many difficult challenges that there was an expectation this would be another example of the nation rising to defeat an enemy.

Our inability to answer seemingly basic questions added to the general frustration. At that contentious congressional hearing, for example, both Tony Fauci and I were asked to predict how many people eventually would be infected and how many would eventually die. There was no answer to that question. Tony responded, "I can't give you a realistic number until we put into the factor how we respond. If we are complacent and don't do really aggressive containment and mitigation, the number could go way up and involve many, many millions. If we start to contain it, we could flatten it."

People wanted a scapegoat. I understood that. This had to be somebody's fault. That initial criticism focused on the tests. We didn't produce enough of them. Our test didn't work. We didn't distribute them widely enough. We were too busy to defend our actions. Accepting criticism was part of the job.

I can remember having a great deal of hope as we struggled to get our footing through January, and while we were creating and circulating tests in February and the first half of March. We were trying to hold back the tide until we had some way to fight back—hoping and praying that maybe, with luck, we could keep the total number of cases in our country under ten thousand. That was a manageable number. I can also remember how saddened I was when it became very clear that wasn't going to be possible. When it was obvious we were going to be dealing with hundreds of thousands of cases, my focus became doing everything possible to prevent us from reaching that model of 2.2 million Americans dead.

If it is possible to point to the moment the pandemic became a national health crisis, this was it. It was as if innumerable creeks, streams, and rivers had flowed together to create a whirlpool. We were deluged with data, reports, questions, demands, requests, and pleas to meet ongoing issues while also trying desperately to minimize the multiplying rate of infection and, unfortunately, death.

Most of the new initiatives I had intended to pursue had to be put aside: the war on drugs, obesity, diabetes. We did as much as possible to fulfill our other national health responsibilities, and everybody's days got considerably longer. But in reality everything other than Covid became a distant second.

The daily reports were never good. By early March, the virus had spread to all fifty states. Every morning, the number of infections would grow. There was never any respite. There was little hope. And every day I would confer with the people trying to create vaccines and identify existing pharmaceuticals that could stem this wave. Mostly though, it felt like we were sticking the proverbial finger in the dike. We constantly were moving our people around the country, sending them wherever the next major problem occurred. In that sense, it was like Whack-A-Mole. We tried to be responsive to requests from state, local, and tribal health departments. There literally were hundreds of outbreaks occurring in nursing homes, meatpacking plants, prisons, schools, and any other place where people gathered or were housed together. The authorities would call and ask for help. In response, the CDC deployed more than three thousand members of our COVID-19 task force to every state in the country so we could maintain a direct line of communication, getting up-to-date data from those places while keeping state health departments informed.

When invited, our people would visit the facility to assess the extent of the outbreak. One person? Ten people? What were the practices that led to a lack of containment? What can be done to stop transmission? We did that in hundreds of facilities, thousands of facilities. It wasn't enough, not nearly enough to do what was becoming necessary. We just didn't have the human capacity to orchestrate an effective response.

We received different levels of cooperation from the various states. Many states welcomed us; others were wary. For example, the data compiled by our people in North Carolina showed an unusually high number of Hispanics were infected. That was unexpected. Why was there such an unusual outbreak spreading throughout primarily Hispanic communities in that state? I wanted to go in and investigate. There might be a clue there that would add to our knowledge. My CDC people there told me they needed permission from the state to do the research. To my surprise, North Carolina turned us down. The state government turned us down.

Later we discovered that a large number of Hispanics living there worked in agriculture, especially chicken and meatpacking plants. They lived in crowded housing and commuted mostly together in crowded vehicles. In that environment, there was almost no way to prevent the disease from spreading.

New York welcomed government assistance. Through most of March, New York remained the epicenter of the pandemic. That was logical. New York City is densely populated and is the nation's leading tourist destination. The pride that the city takes in being America's "melting pot" really worked against it. Until the travel bans went into effect, people from around the world, some of whom likely were infected, visited the popular attractions. The first case in the state was confirmed on March 1 when a resident who recently had returned from Iran tested positive. Two weeks later, New York City Mayor Bill de Blasio "temporarily" shut down the city's public school system, the largest public school system in the country, and days later remote learning began. Nothing helped. On the 27th, the state reported 23,000 cases and 365 deaths. Ten days later, the state reported 72,000 cases and at least 2,500 deaths.

It was a harbinger of what the rest of the country was going to confront. The state's healthcare industry was overwhelmed; its hospitals were packed so completely that patients on stretchers lined the hallways, while lobbies and conference rooms were being transformed into wards. Morgues and funeral rooms were running out of space to store bodies. To offer assistance, the government sent a hospital ship, the USNS *Comfort*, to the city to relieve some of the overcrowding by initially treating patients needing help for conditions other than Covid. It was a good idea that didn't work. By then, so many people were self-isolating that there were far fewer accident victims needing assistance, so the ship remained mostly empty. The Defense Department then authorized the *Comfort* to accept Covid patients.

Cuomo also arranged with the Defense Department to transform the city's Javits Convention Center into a 2,500-bed hospital; like the ship, it initially was scheduled to treat non-Covid patients, but almost immediately it became obvious it was needed to care for people suffering from the virus.

But perhaps the most significant issue centered on nursing home residents. As we had learned, elderly people were by far the most susceptible to this virus, especially those with existing health issues. The type of people living in healthcare facilities. Many of them had to be hospitalized. As the hospitals rapidly became overcrowded, the question became where should these patients go after recovering. There was a lot of debate about that. The fear was they would be readmitted to nursing homes and spread the virus to the at-risk population.

I had several discussions about this with both Governor Cuomo and Harold Zucker. Harold and I discussed making some of these facilities Covid Only. I suggested they send all of their recovering patients to a single nursing home. There were some closed facilities that could have been used. I also urged them to protect those places by prohibiting visitors and making sure the staff was tested frequently. I had those same discussions with governors of several other states, including Florida's Ron DeSantis

When those discussions began, we still believed we were dealing with a SARS-like virus; if you were infected, you had to be symptomatic. Therefore, after the symptoms disappeared, we thought people were no longer infectious. On March 13, we issued a directive reading in part "nursing homes should admit any individuals from hospitals where Covid is present." We also issued strict guidelines that before being allowed back in those homes, patients should be tested to ensure they were not a danger to others—and, even then, they should be isolated for fourteen days.

As I understand it, New York put people back in nursing homes without testing them. It is impossible to successfully isolate people if you don't know who is infected and who isn't. With everything going on during that period, I wouldn't blame anyone for decisions that were made. Too many decisions had to be made too quickly without sufficient information.

However, Governor Cuomo did blame the CDC, telling reporters, "I just want to reiterate once again that the policy . . . was in line directly with the March 13 directive put out by the CDC."

Well, it wasn't. We never recommended people be returned to nursing homes without being tested and isolated. We never suggested just close your eyes and put them in a nursing home. Our directive advised that patients might be discharged from a hospital and admitted to a nursing facility "only if the nursing home can implement all recommended infection control procedures." Admittedly, I was extremely disappointed when he made those comments.

The result was disastrous: thousands of deaths. As many as 40 percent of the infections and subsequent deaths in New York began in nursing homes. Who was responsible for that became a highly charged political dispute, exactly the type of situation neither I nor my people had time for. In my opinion, while allowing asymptomatic patients to return to a nursing home without being tested contributed significantly to the horrendous total, that wasn't the primary cause of those infections. I know some people don't like to admit it, but a large number of infections were brought into those homes by well-meaning members of the staff, particularly nurse's aides who worked in multiple nursing homes or hospitals. Once they carried it into that world of vulnerable people, it couldn't be contained.

In the ensuing years, several reports would be issued trying to assess responsibility. Unsurprisingly, they reached a variety of different conclusions based mostly on the objectives of the organization issuing them. The reality was that it took a confluence of events during an extraordinarily hectic period, combined with a lack of scientific information, to create that crisis. In retrospect though, I'm not certain there was much more we could have done at that time to limit the terrible toll.

The monster was destroying the city. That city was New York. We had lost the battle to stop it, to contain it. As the death toll climbed, it felt like this monster had the world in the palm of its hand. All we could do was try to save as many lives as possible.

FIVE

We had lost the first battle. We had failed to contain the spread of the disease. COVID-19 was spreading throughout the country, spreading from the urban areas where it first took hold to the small rural areas, and there was little we could do to stop it. No one was safe from it. Our goal then became trying to protect people. As the New York City Public Health Department had written during 1929's flu season, "Our publicity policy this Winter, which has been to educate the public . . . that sensible preventive measures on the part of the public could do more than anything else to keep the disease from spreading had been a great help in combating influenza. Since medical science has been unable thus far to evolve an effective preventive or cure for influenza, the most effective way of fighting the disease has proved to be to educate the public to take no chances and go to bed and call a doctor at the first sign of influenza."

Almost a century later not much had changed. In the early years of the AIDS epidemic, for example, there were all kinds of suggestions and discussions about how to keep people safe. Many of them were sensible, others were ridiculous. The problem then, as it has been throughout history, is that people felt vulnerable and were frightened. Since it was established that HIV could be spread by infected blood, experts decided healthcare professionals could protect themselves by wearing plastic gloves.

Maybe there was some benefit from doing that, but what it really did was create a false sense of safety. Medical personnel didn't get AIDS from getting blood on their hands. They got HIV from being stuck by a needle or cut by a scalpel which allowed infected blood to break into their body. It made a lot more sense, I suggested, to use retractable needles, shatterproof tubes, and other materials that would prevent that. But too often, misinformation drove responses.

I remember driving home one night during that pandemic, for example, when I came upon a car accident. People were hurt, and they were bleeding. I stopped and gave them first aid until the police and ambulance arrived. As a result, I had a considerable amount of blood on my clothes. When my ten-year-old daughter saw me, she got very upset. She began lecturing me, telling me that her pediatrician always wears gloves when she sees him, and I needed to wear gloves too, or I might get AIDS. I explained, "Jenny, that's not really how you get AIDS." I explained that even when there was blood, there had to be some kind of skin wound for the virus to be transmitted. My son went into the kitchen and got a pair of Joy's latex gloves, and he sliced them with a knife to show his sister how little protection they actually provided.

The fact is that this was not a well-thought-out policy. Gloves might decrease hepatitis B or C transmission, but they had little impact on HIV. The problem with that policy was that it sent a mixed message to society. We were telling teachers that it was safe to have HIV-positive kids in their classrooms, yet they saw images of healthcare professionals wearing gloves to treat patients.

When I lectured about AIDS, I spent time talking about the use of condoms. "They do provide some protection," I explained, "but not absolute protection." I went through which types and brands were the most reliable, and I discussed jellies. "But there is one method," I went on, "guaranteed to offer complete protection." As I concluded my presentation, I showed a slide of a man wearing a leisure suit sitting at a bar, a sly smile on his face, with the caption reading, "Hey babe, you wanna drink?" I concluded, "Leisure suits offer a lot more protection than condoms."

We had no effective medical treatment for AIDS, so we proposed basic, sensible behaviors that would decrease the chance of contracting the disease. We were in a similar situation in the first months of 2020. Work was progressing on a vaccine, but we had no idea when it might be available, and until that time the reality was that we had to rely on updated versions of those same methods people had used throughout history to protect themselves from infectious diseases.

Societies have been searching for that protection for thousands of years. By the fourth century BCE, the Athenian Thucydides, often considered "the father of scientific history," acknowledged that prayers to the gods had failed to avert the plague, so citizens had to find other means to control the disease. The most effective has always been staying away from infected people or environments. Long before science discovered the existence of viruses, it was known that in some way diseased people and objects could spread disease to others. In response to the Black Plague, for example, a "cordon sanitaire" isolated the entire Italian peninsula and the border between the Austrian Empire and Turkey. Gunboats forced ships to drop anchor offshore or dock at carefully guarded designated ports. Soldiers and police guarded the shoreline and beaches. Those people allowed to land were isolated in makeshift camps. When a ship arrived from a plague-infested area, its captain was brought to shore in a lifeboat, and then he was put in a small enclosure to speak with local health officials several feet away through a small window. He had to present proof that his crew and passengers were healthy and whatever cargo he carried was not infected. If there still was suspicion, the ship was directed to a quarantine station, where it was isolated and fumigated for forty days.

Similar sanitary cordons, enforced by armed guards, were positioned around the city-states, and strangers were prevented from entering—by force, if necessary. Through separation and isolation, the threat of the plague and other transmittable diseases was diminished and even in certain places eliminated. That concept of keeping a safe distance was often continued after death. In several places, the possessions of victims—and sometimes even their homes—were burned.

In 1793, a yellow fever epidemic swept through Philadelphia killing as much as 10 percent of the city's population. Even the traditional methods of prevention, including burning sulfur and urine-soaked straw in smudge pots, did not slow its ravages. A year later, desperate citizens set up a permanent Board of Health and granted it authority to establish maritime quarantines and enforce sanitary regulations.

Preventing contact by isolating patients remained the best weapon. Around the world, countries were building hospitals restricted to patients suffering from diseases like smallpox and leprosy. Other so-called "fever hospitals" were built for patients with diphtheria, scarlet fever, yellow fever, and even cholera.

Gradually, medical science began to understand that sanitation—including actions as simple as washing your hands—also could play an important role in preventing the spread of disease, even if they didn't know why. In her classic 1860 *Notes on Nursing*, Florence Nightingale wrote "Every nurse ought to be careful to wash her hands very frequently during the day. If her face, too, so much the better." She advised women to help keep their family healthy in their home by opening windows to increase ventilation while ridding the house of "stagnant, musty and corrupt" air. Also, she cautioned, the proper mistress should clean "every hole and corner of the home" because dirty carpets and furniture "pollute the air just as much as if there were a dung heap in the basement."

We were learning slowly, and mostly through trial and error rather than scientific studies, but we were learning. Or so we believed. In 1918, all those lessons proved worthless. It is generally accepted that the Spanish flu, the influenza pandemic that killed as many as 50 million people worldwide and 675,000 in America, actually started here. On March 11, one of the first known cases emerged at Fort Riley, Kansas. As the army rapidly mobilized to go to war, it became the perfect breeding ground for an infectious disease. Soldiers were crammed together in often unsanitary barracks, shared equipment and meals, then sailed to Europe on densely packed troopships.

The American Expeditionary Force brought with it courage, spirit, equipment, and a deadly virus. It spread rapidly through all the armies of the Great War. Americans called it the "three-day fever" or the "purple death" because the victim's feet sometimes turned a dark color. To the French, it was "purulent bronchitis." Italians called it "sand fly fever." And Germans complained of "Flanders fever." Returning troops brought it home with them, infecting others who carried it around the world. It took only four months for it to infect the entire world, from the great continents to the small seemingly isolated islands.

Wartime censorship in most nations had prevented newspapers from reporting the scope of the devastation, claiming that releasing that information could benefit the enemy. So, few people knew that millions more people were being killed by the flu than were dying in the war. But because Spanish newspapers openly reported the toll of the pandemic, it became known as "the Spanish flu" and "the Spanish lady."

By the winter of 1918, it was killing thousands of Americans every day. In cities throughout the country, public gatherings were prohibited; schools, dance halls, and pool halls were closed; church services were canceled; and sneezing or coughing in public was made illegal. People began wearing gauze masks, initially making them at home. But soon, community groups, churches, civic organizations, the Red Cross, and local public health groups were holding mask-making sessions—during which, naturally, they all wore masks. Eventually, as companies recognized the potential for profit, they were mass produced. Many cities passed regulations making it mandatory that people wear masks on public transportation. A *New York Tribune* sportswriter previewing a boxing match suggested, "Perhaps Joe Jeannette and Kid Norfolk will enter the ring wearing the latest thing in influenza masks, and perhaps instead of handing out rain checks to ticket holders in case of a postponement the management will distribute 'sneeze checks.'"

New York City created a squad—called the "sanitary police"—whose mission was to enforce public health measures such as wearing a mask, social distancing, and otherwise following quarantine restrictions. They had orders to ticket or even arrest people who spit, sneezed, or coughed in public. There was nothing humorous about it. People were dying from those violations. In particular, regulations against spitting in public were strictly enforced. Posters were displayed in public areas reminding people to cover their mouth and nose when sneezing or coughing. People caught spitting on subways, elevated trains, or in public spaces were arrested and taken to court. As the *Tribune* reported in October, "The Health Department's campaign against 'spitters' in public places resulted in five hundred arrests during the week. Sanitary police are closing motion picture theatres, saloons, and soda fountains daily." Spitters were usually fined one dollar, which actually was a meaningful penalty then.

While there were pockets of resistance to the different restrictions in different cities, there was no unified movement as the nation would see a century later. This was before cell phones and the internet. Limited communications made organizing widespread protests difficult.

Helpless, frightened, and desperate, Americans were willing to try all types of remedies including consuming commonly used treatments like aspirin and alcohol, rubbing raw onions on their bodies, camphor oil, enemas, and laxatives. Actually, the use of raw onions made some sense—it certainly would encourage people to stay away from you. They also turned to other kinds of strange remedies, among them beef gravy, bloodletting, and gargling saltwater (because salt was known for its antiseptic properties). Ultimately, they turned to the biggest snake oil of all—actual snake oil.

Among the most unusual proposed treatments was the inhalation of noxious fumes. This was based on anecdotal "evidence." Supposedly, rumors spread that English factory workers who were exposed to noxious gases in their work had significantly lower death rates than other people. The story spread that in one town the infection rate among the general population was 40 percent while for men who worked with nitric acid it was only 11 percent. That was reduced to 5 percent for those who worked around gunpowder. As a result, British parents were taking sick children to the local gasworks so they could inhale fumes.

Among those noxious gases was chlorine, which was used as a disinfectant. In fact, in large doses chlorine can kill both germs and viruses—but the amount it takes to do that makes it poisonous.

All the various methods known to be effective in combating the Spanish flu were summed up in a release from New York Health Commissioner Royal Copeland:

- Keep away from the cougher, sneezer or spitter who does not use a handkerchief.
- Keep out of crowds whenever possible.
- Don't use dishes or towels which have been used by others until they have been washed in boiling water.

- Don't put your lips against the telephone mouthpiece and don't put into your mouth a pencil or any other article that has been used by another.
- Sleep in a well-ventilated room under plenty of bed clothes.
- Walk instead of using the street car or subway whenever your journey is a short one.
- Be temperate in eating and observe the ordinary rules of hygiene.
- Wash your hands and face immediately upon reaching your home and change your cloths if possible before mingling with the rest of the family.

Copeland also offered advice about treating infected people:

- Go to bed upon the first indication of illness and call a doctor.
- The sick person should have a room by himself.
- Care should be taken to have the sick person cough, sneeze or expectorate in gauze, which should be burned at once. Persons handling this gauze should wash their hands after each attention.
- Patent medicines should be avoided.
- The patient's room should be well-ventilated; care should be taken that no draft strikes him.
- Visitors should be kept from the sickroom.
- The patient should remain in bed long enough after the fever in influenza has subsided so that he will no longer be in danger of an attack of pneumonia.

Literally 101 years later, those guidelines were not substantially different than what the CDC suggested. Our initial guidelines were probably based more on common sense than scientific data. We just didn't know enough about this virus. The best thing people could do to avoid getting sick was to stay away from sick people. We recommended maintaining a six-foot distance between people. That quickly became known as "social distancing." This was a relatively new phrase. It was probably used for first time in 2003 during the SARS epidemic. Then, a few years later when the avian flu was threatening to spread, the *New York Times* wrote, "If the avian flu goes pandemic while Tamiflu and vaccines are still in short supply, experts say, the only protection most Americans will have is 'social distancing,' which is the new politically correct way of saying 'quarantine.'"

That distance, six feet, was a compromise. I was not in favor of it. I thought that for a variety of reasons keeping at least three feet apart and washing your hands often would be fine. But I lost that debate. It wasn't' an "inexact science," it was nonscientific. Some knowledgeable people at the CDC suggested a distance of as much as ten feet, which the administration sensibly rejected as "inoperable." It became an issue when Tony Fauci told a congressional committee that distance "just appeared," suggesting there was no scientific data to support it. There certainly wasn't much, but his comment was misunderstood. It made it seem like we were just making it up. He later clarified his comments, explaining "What I mean by no science behind it was that there wasn't a controlled trial that said compare 6' with 3' with 10', so there was no scientific evaluation." That distance, he continued, was based "on studies years ago that showed that when you're dealing with droplets, which at the time the CDC made that recommendation it was felt that transmission was primarily through droplet, not aerosol (tiny particles that can be spread a greater distance in a mist-like spray), which is incorrect because we know now aerosol does play a role."

He concluded his testimony by reminding Congress, "That's just one of the things that got distorted."

There was some reasoning behind that distance. Obviously the farther away a person is from other people the risk of being infected drops because the droplets, or the spray, get diluted by the air. Droplets actually fall to the floor relatively quickly. As I learned, in 1897 German bacteriologist Carl Flügge proposed maintaining a six-foot distance to prevent the spread of germs. Decades later, after the invention of high-speed photography scientists confirmed that talking, sneezing, coughing, or yelling could propel particles as much as six feet. Spit, obviously, could go much farther.

There were very few scientific trials. In a 1948 study, researchers set petri dishes at various distances away from a subject with a bacterial infection.

The World Health Organization disagreed with our recommendation, issuing guidelines that three feet was a safe distance. But in this country, people adapted to the realities of six-foot social distancing. Post offices, shops, and most places where lines form placed "safe circles" at the correct distance, some of them doubling as advertisements. Public parks drew chalk circles on the grass so people could sunbathe safely. Once it was determined that two open umbrellas created almost six feet of distance, people began using their umbrellas as shields. Rather than traditional greetings like shaking hands or even hugging, people touched elbows and learned how to sneeze into the crook of their elbows. One priest in Michigan dispensed holy water by having his congregants drive past him while he sprayed them from a distance with a water gun. A Maryland events planner offered round tables surrounded by inner tubes that prevented diners, who sat in the middle, from getting too close.

Initially, at least, some officials in the White House supported these suggestions. Deputy National Security Advisor Matthew Pottinger and half his staff moved out of the West Wing into an isolated office in the Eisenhower Executive Office Building. He also basically stopped meeting face-to-face with his boss, Robert O'Brien, just in case one of them got infected. Provisions were made for other employees to maintain a working distance.

People adjusted. Actually, some people adjusted. There were a lot of people who ignored these suggestions. Various media outlets and a lot of people challenged the six-foot concept. Having fought the AIDS wars against misinformation, criticism, personal attacks, politicians, bigots, quacks, and self-promoters, I was not the slightest bit surprised. There is a long history of Americans speaking out or taking action against public health advisories.

What became known as the anticontagionist movement began in the United States during the 1793 yellow fever epidemic. Anticontagionists believed that infectious diseases were not spread by contact with people or materials but rather were caused by unknown environmental factors or just general miasma. These diseases, they believed, actually resulted from poor sanitation, overcrowding, contaminated water, breathing air fouled by decaying organic matter, or even germs found in soil. As a result, they argued against strict quarantines and isolation, complaining that those restrictions prevented travel and free trade, severely damaging the economy.

The leader of the anticontagionists was Dr. Charles Caldwell, who believed quarantines were a historic relic from "the dark ages, when Europe was a stranger to physical science," that they were "erroneous, destructive," and "a false idol founded on superstition and prejudice" that created "bigotry and delusion" and should be entirely abolished. Instead of quarantines, he advocated free trade without burdensome detentions, fighting disease by preventing overcrowding and improving waste disposal, drainage, ventilation, and personal hygiene.

After making an impact on public health policy, the anticontagionists gradually evolved into a more general health freedom movement—a mixture of politics, skeptics, mistrust in government or authority, and the fundamental right to make personal choices about your own healthcare.

New Englander Sam Thompson became rich and famous by challenging accepted medical practices; in the early 1800s, he denounced conventional medical treatments and instead patented his own herbal remedies and built a nationwide franchise system to sell them. His supporters, Thomsonians, justified their belief by quoting the probably fictitious Honestus, who supposedly said, "If I be conscientiously opposed to bleeding, blistering, mercurializing or poisoning with emetic tartar, opium, arsenic or prussic acid shall I be compelled to employ a law-made doctor, who deals almost exclusively in these potent remedies." By 1839, he claimed to have three million followers who felt strongly that every American had the freedom to choose what treatments to accept to protect their own physical well-being.

New York's Friendly Botanic Society spoke for a substantial segment of the American population when it resolved in the 1840s that all people should have the "freedom to choose the means which we believe are best calculated to secure us health and life."

The grand tradition of resisting government mandates was embedded in the fabric of this country. There have always been people willing to remind us of that. That resistance has surfaced throughout the history of public health in America. In every health crisis, there has always been a percentage of people who refused to adhere to suggested guidelines, claiming they were an infringement of their personal freedom. During the 1918 pandemic, for example, San Francisco passed an ordinance compelling people to wear masks in public. "Wear a Mask and Save Your Life!" the Red Cross urged, claiming without evidence "A Mask is 99% Proof Against Influenza." In response, thousands of people joined the Anti-Mask League, which argued these regulations were unconstitutional. The city responded by arresting hundreds of people for "disturbing the peace" and sentenced them to ten days in prison or a $5 fine.

So we knew what to expect when we began actively discussing limiting social interactions by urging states to shut down schools, shops, businesses, and other venues where people gathered. It was a mammoth decision for governors and mayors. Lives and careers were going to be changed by whatever decisions were made, and there was very little scientific evidence to guide us. It was another precarious balancing act. We certainly were aware of how unpopular that decision was going to be. But I couldn't help thinking about that model, 2.2 million dead Americans by September if we didn't take aggressive steps. This was not our decision, this was up to each governor or mayor, but I participated in meetings and as CDC director offered my opinion.

Whether or not to shut public schools probably was the first dramatic decision most American governors and mayors faced. The CDC was releasing statements every day trying to keep people informed. In late February, while we were still hopeful, we could prevent the general spread of the disease, we advised, "Community-level nonpharmaceutical intervention might include school dismissals and social distancing in other settings (postponement or cancellation of mass gatherings and telework and remote-meetings in workplaces.) . . . Studies have shown that early layered implementation of these interventions can reduce the community spread. . . . These measures might be critical to avert widespread COVID-19 transmission."

By mid-March, as the virus was raging across the country, New York, Los Angeles, and Seattle had shut down their schools. "This is not something in a million years I could have imagined having to do," a distraught New York Mayor Bill de Blasio said, later adding that it was an "extraordinarily painful" decision.

The rationale was that children sitting in crowded classrooms and walking in hallways would be very susceptible to the virus and then would become spreaders, taking it home to their family. I understood that; I just didn't agree with it.

It was an extremely controversial decision, especially among parents with young children who relied on schools to keep their kids safe while they were at work. There were strong arguments to be made on either side, and there was very little historical precedent one way or the other. New York City schools, for example, had remained open during the Spanish flu pandemic. Their school buildings had been designed starting in 1898 with large windows that provided both light and ventilation. The schools also served the city's large immigrant population with a robust public health program, which included the nation's first school health inspectors and school nurses. Students were routinely inspected for signs of a respiratory infection and if they had any symptoms were isolated, sent home or to a hospital. So students actually were safer in those buildings than at home in crowded tenements.

As it turned out, a majority of districts around the country did close their schools in 1918 for periods ranging from a few days to months, which provoked substantial anger. After Minneapolis shut its schools, for example, the state health officer objected, asking sarcastically, "Do you think that any program of shutting up a few things is going to stop this epidemic?" A week later the Board of Education voted to reopen the schools, and that city's chief public health officer instructed the police department to arrest the board members. The board diplomatically reversed its decision and closed the schools. But there was never any data that showed a significant difference in the health outcome if schools were kept open or closed. A 1940s study of the actions taken during the Spanish flu reported that school closures during that pandemic did not have a long-term impact on "attendance, educational attainment or adult labor market outcomes."

The little historical evidence we did have was all over the place. After an estimated 100,000 Americans died in the 1957–1958 Asian flu pandemic, another study showed that school closures supposedly reduced morbidity by 90 percent. Data compiled between 2004 and 2008 indicated that closing schools cut down infections by as much as 50 percent. When swine flu spread throughout the world in 2009, Mexico shut its schools and advised people to avoid greeting others by shaking hands or kissing, which reduced transmission by as much as 37 percent. An article published that year in Britain's *Lancet Infectious Diseases* concluded that closing schools interrupted the course of the infection and slowed its spread, giving scientists time to create a vaccine. But the study did not reach any conclusion about the potential impact of the school closures on the future lives of those students.

For all those reasons, on March 13, the same day we announced we would be closing the cruise industry, we issued an advisory, which was about all we could do, suggesting—suggesting—that closing schools for as long as eight weeks "might" be an effective way to contain the virus. To facilitate that, the Department of Education told local officials that "temporarily dismissing childcare programs and K–12 schools is a strategy to stop or slow the further spread of COVID-19 in communities" and waived several regulations required by federal law. By that time though, more than 21,000 schools had already been closed by local authorities, and a significant number of colleges and universities had locked down their campuses and sent students home.

There were more than 50 million kids in public schools, and I was against closing them. But I was in the minority. I had read the studies that existed about previous infectious disease epidemics; in fact, I'd lived through several of them. I felt the advances we had made in improving ventilation in schools made them considerably safer than they were during previous outbreaks. So I thought it was a big mistake to move away from face-to-face learning. I thought the public health interest, kindergarten through twelfth grade, was better served by keeping schools open. There were a lot of reasons for this. The initial evidence we were compiling seemed to confirm that young people, unlike during the Spanish flu, actually were in little danger from COVID-19. While the guideline for social distancing was six feet, I believed a distance of three feet was sufficient to prevent infection; with some management, that could be maintained in classrooms and hallways. If kids stayed a normal distance apart and washed their hands, and if they used masks when advisable, the benefits in keeping schools open were obvious.

It also seemed likely to me that virtual learning could not be as effective as being in a classroom. I also thought it would lead to some isolation and mental health issues, such as depression. I thought it could even lead to drug abuse. And then there is the fact that the significant number of kids who rely on their school for breakfast and lunch would be deprived of healthy meals. In addition to educating young people, America's public school system provides an array of physical and mental health care as well as vital social services.

The truth is we didn't know what the outcome of closing the schools would be. Nobody did. There was no way of predicting it. One thing we did know is that it would cause serious problems. As the AP reported about our advisory, "The CDC conceded that long-term closing could significantly affect academic outcomes for students, economic conditions for families and health conditions for grandparents who care for students."

For me, the key to keeping schools open was providing good air quality. Making sure the kids were in a highly ventilated space. Moving air significantly cuts down on infection. I felt so strongly about this that the government offered school districts billions of dollars to purchase air purifiers. These funds could be used for a variety of purposes, we agreed, and among them was "inspection, testing, maintenance, repair, replacement, and upgrade projects to improve the indoor air quality in school facilities."

Unfortunately, for whatever reasons, a majority of districts did not take advantage of these funds.

Any efforts to keep the schools open collapsed when President Trump issued his recommendation on March 16, "My administration is recommending that all Americans, including the young and healthy, work to engage in schooling from home when possible, avoid gathering in groups of more than ten people, avoid discretionary travel and avoid eating and drinking at bars, restaurants and public food courts."

While the White House announced this advisory would be reviewed in fifteen days to respond to the existing situation, it warned this emergency could stretch into the summer. I think a fundamental problem we had to deal with was an American public that had been conditioned to expect straightforward answers and direction, which we just couldn't provide. People wanted to be told, do this and you and your family will be safe, but we couldn't do that. We didn't have those answers.

I'll tell you who did: people on the internet. One of the things that made dealing with this pandemic so different and difficult from previous public health issues was the World Wide Web. This was really the first time anyone had to deal with a spreading disease since the creation of the internet. In theory, it is a remarkable tool: almost every American has instant access to information on their computer or mobile device. The reality is different. While the benefits are obvious, there are no barriers to prevent inaccurate, misleading, or fabricated information from being posted and circulated widely. And there are people who will believe almost anything that shows up on their feed, especially if it fits their preconceived skepticism—psychologists call it confirmation bias—and they will connect to other people who feel the same way.

If the anticontagionists or Thomsonians had access to the internet, they might still exist. We were dealing with their natural descendants.

We expected pushback, and we were never disappointed. Whatever advisory we issued was almost always met with vocal opposition. Some of it was based on different legitimate interpretations of data and history, but some of it was simply created by self-proclaimed "experts" to advance an individual or an organization. At least some of these resistors asked for contributions to organize and spread the word. I don't know for certain, but I suspect a lot of money changed hands. The internet allowed dissenters to find each other and bond into a sizeable resistance. That was a new phenomenon for a public health agency. I didn't have the time nor the interest to take in the opinions of these people, so I rarely read their comments. Unlike these people, I'd spent my life in this field, and I knew very well what we did and didn't know.

Until this fifteen-day pause was announced, this was primarily a public health issue. At first, I didn't consider it a shutdown, but rather a last-ditch attempt—an aggressive mitigation to try to slow the spread of the disease. At that time, we were beginning to see the collapse of the public healthcare systems in New York, Detroit, Chicago, Los Angeles, New Orleans and other cities. In New York, the mortality rate was 8 percent, which put us on the curve predicted by that ominous model. But I did not support shutting down the country. I advocated a pause, which would give us time to understand what was going on and figure out how to keep our schools and economy open in a safe and responsible way.

The only type of business that I came down hard on was crowded bars, those places where fifty, one hundred, or more people were packed together, shouting to be heard above the music, spraying particles every time they opened their mouth. I just didn't see any steps these places could take to suddenly become safe. But I did feel strongly that we could find ways to mitigate the danger in most other aspects of our economy. That there was no reason for a complete shutdown of the country.

But that initial fifteen-day pause transformed the pandemic from a purely public health issue into a critical economic issue, which made it the top-notch political issue the president was facing. I continued to argue that the key to protecting our economy was first and foremost to take the right steps from a public health point of view, but that's not what some of the administration's economic advisors believed. The political pressure they put on us made it harder to keep the focus on the best public health answer, rather than what was good for the administration. They made a strong effort to control comments, reviews, and public statements. There were multiple times where they applied a lot of pressure.

The administration made Dr. Scott Atlas its point man on the task force. This was really when our response got taken out of the hands of epidemiologists and became more politically sensitive. Scott Atlas was a radiologist and knew little about infectious diseases. But he was an astute politician. He and I disagreed about a lot of things. For example, he convinced the president and several members of the cabinet that we didn't need to continue testing to identify people who were asymptomatically infected, that we could modify our guidance. I didn't agree. But in our task force meetings, we forged a compromise that I could still accept, vague language that the public health community could live with. Basically, it advised people to take logical steps: minimize exposure, maintain social distance in public places, and practice basic hygiene. And individuals experiencing any potential symptoms should self-isolate and get tested as soon as possible. What I wanted to do was encourage people to have a serious discussion with their physician. I believed our guidance encouraged the public health sector to continue what they were doing: testing, diagnosing, isolating, and following recommended procedures.

Within forty-eight hours, it became clear to me that the public was misinterpreting our guidance to conclude they didn't need to be tested. The way it was packaged it seemed to be informing healthcare professionals it was no longer recommended they diagnose and isolate asymptomatic people. That we didn't need to be concerned about this "silent epidemic." That it wasn't necessary to find those people and isolate them. That was wrong; it couldn't have been more wrong, and I was angry with myself for accepting it. I immediately revised the guidance and published it. That created a big argument in the task force. Why did I change it? I did not have the task force's approval to change it. Atlas, in particular, was very aggressive, ripping me apart in front of Vice President Pence. He might have even gone to the president to protest that he had not agreed to the change in language. It got very contentious. There was considerable pressure on me to change my guidelines. For three or four days I was criticized by both the task force and the White House. There was some yelling, especially when I told him, "Scott, this isn't the task force guidance. This is the CDC, and you don't have a role in approving CDC guidance. That's the job of the CDC director."

I slept peacefully every night. I was content that I never put anything in front of public health principles and outcomes—with that one exception. And I learned from that.

There were bound to be disagreements. Never before in history had people tried to deal with something like this with the technological tools we now had available. There was no one size fits all. Some loud discussions took place behind closed doors as we tried to maintain the image of a unified task force. We weren't. But I had deep respect for the opinions of Birx, Fauci, and several others. When we disagreed, we tried to find a balance between our competing points of view. These were people who were concerned about keeping people alive, not the next election. But we had to deal with reality. Had it been up to me, the messaging would have been clear, as opposed to being cloudy or even contradictory. It would have been more definitive. I'm not claiming I had the right answers, we still were deep into the learning mode, but I can state unequivocally that my public health guidance would not have been tempered by political considerations.

It seemed obvious to me that Atlas had a different agenda. He was representing an administration running for reelection, and I accepted that even when I disagreed with him. The situation had begun deteriorating soon after he joined the task force. Atlas was skilled at exploiting any differences that surfaced between us. Eventually, to prevent that from happening, several of us began meeting privately. Tony, Deborah, FDA Commissioner Steve Hahn, Surgeon General Jerome Adams, a few other healthcare professionals, and I would get together before the task force met to discuss whatever disagreements we had, so we could present a somewhat unified position.

This was the beginning of the political realities that we had to deal with throughout the pandemic. Neither Atlas nor HHS Secretary Azar wanted to take the chance I would contradict their narrative, so essentially, I was prevented from communicating directly with the public. In fact, the administration barred Tony Fauci and me from appearing in front of committees in the Democratically controlled House of Representatives.

I finally learned I could get my message out by participating either in person or virtually in town halls and other informational events. I also met with the local press, which was thrilled to be able to interview the director of the CDC.

The fifteen-day pause gave us the opportunity to take one brief safe breath. We knew that we were just buying time, maybe saving some lives, until we had a vaccine. The vaccine was the Holy Grail. It was going to make all the difference.

We believed that.

SIX

I have always believed that vaccines are the most important gift science has given to society. While British scientist Edward Jenner is credited with developing the first practical vaccine in 1796, when he inoculated an eight-year-old boy with cowpox and demonstrated it would protect him from the much more serious disease, smallpox, the concept that exposing a healthy person to a mild dose of a disease could provoke natural immunity actually was known hundreds of years earlier.

"Variolation," as it was then known was already being practiced in China and India by the 1400s. Chinese practitioners used a pipe to blow dried smallpox scabs into a patient's nasal passage, stimulating a then mysterious response that provided protection against developing a full-blown case of the disease. It apparently was popularized in America by 1721 by the religious leader Cotton Mather, who learned about it from his slave Onesimus, who had been inoculated in Africa.

The treatment was considered extremely controversial. In fact, because it required introducing live virus into a patient's bloodstream it was risky and caused some seemingly healthy people to develop full-blown smallpox and become infectious. No one understood how or why it worked. Many people who supported it believed it was a gift from God, while those who opposed it claimed it came from the devil.

Benjamin Franklin was a strong advocate. When a smallpox epidemic began spreading in Philadelphia in 1731, he informed readers of his newspaper, the *Pennsylvania Gazette*, that inoculations had saved the lives of many Bostonians when the disease struck that city a year earlier, urging them to ignore the "anti-variolationists." "The practice of inoculation for the smallpox begins to grow among us," he wrote. "How groundless all those extravagant reports are, that have been spread through the province to the contrary."

Several years later, his son Francis died from the disease. Franklin had chosen not to inoculate him because he was already weak from dysentery, a decision he regretted "bitterly" for the rest of his life. When rumors spread that Francis actually had been inoculated—and died from the treatment—Franklin wrote angrily that his son "was not inoculated, but received the distemper in the common way of infection." "Inoculation," he continued, "was a safe and beneficial practice."

Edward Jenner's vaccine essentially ended the use of variolation, which was outlawed in England in 1842.

Vaccines are not miracle drugs. They are better than that. They are grounded in science. We now know how they work and why they work. Throughout my career, I have seen their benefits. I've seen the development of vaccines for serious diseases like hepatitis B and HPV. And I knew COVID-19 would not be eradicated until we had a biological countermeasure known as a vaccine. I knew how difficult it would be to create. After decades of research, we still have not been able to produce a vaccine for HIV, and a new flu vaccine has to be created every season—and even then only has a limited effectiveness. My fear was that creating a COVID-19 vaccine would take years.

I was wrong. Fortunately, I couldn't have been more wrong.

It took literally a few hours for a Turkish scientist raised and living in Germany to produce the first candidate, but it had taken other scientists decades to produce the messenger RNA platform on which the vaccine was built. Very very basically, an mRNA molecule contains short strands of genetic material that triggers cells into producing proteins that cause the immune system to develop antibodies. It's an all-chemical version of putting smallpox scabs into nasal passages—it activates the immune system. It turns your body into a drug-making factory, producing the levels needed to fight the infection. Scientists had been studying mRNA for decades; they believed it had tremendous potential for a variety of applications. The problem was they couldn't figure out how to get in into cells. It is so fragile that enzymes easily ripped it into useless pieces.

My father had been trying to master mRNA when he was working at the NIH in the 1940s. I had always been intrigued by it. Early in my career, I'd tried to work with it quite often in labs; I wanted to see the impact animal RNA had on breast cancer viruses, but those experiments failed because I couldn't get it to stabilize. I had come to believe that it simply was too brittle to ever become a valuable tool.

Finally, in 2018, the FDA approved the use of lipid nanoparticles—lipid is a compound that is insoluble in water that is found in cell membranes—as a shell to protect mRNA. Think of it as an envelope strong enough to protect whatever is put inside it. Scientist Uğur Şahin and his wife Özlem Türeci had cofounded the biotechnology company BioNTeach in Mainz, Germany in 2008 to create immunotherapies and vaccines using mRNA to fight cancer and infectious diseases. It was a medium-sized company, and I had never heard of it. That was not surprising; with drugs capable of generating billions of dollars in profits, there are literally hundreds of companies like this around the world essentially mining for biotech gold.

And BioNTech was perfectly positioned when the virus emerged from Wuhan. When the Chinese revealed the virus' genetic sequence in early January, he decided to try to develop a vaccine. Over a weekend he designed ten potential vaccines on his computer. These candidates took a small piece of mRNA from that part of the coronavirus targeted by antibodies and wrapped it in lipid nanoparticles. In theory, that would stimulate the immune system. BioNTech had been working with Pfizer since 2018, trying to develop a flu vaccine, and those two companies formed a partnership. In early April, BioNTech began testing the selected vaccine candidate in a two-hundred-person study in Germany, while Pfizer began trials here with 360 people.

The CDC did not get involved in producing a vaccine; that was the job of the NIH and the private sector. But from day one, we had a front-row seat to vaccine development. I know the NIH started working on a vaccine as early as January 11. Other companies had also been playing around with mRNA, using the new technology to try to create a SARS or MERS vaccine. So they quickly directed their research to this new virus.

No vaccine had ever been produced so rapidly. At the same time, numerous other companies were also racing to produce a vaccine. Millions of lives and billions of dollars were at stake. In late April, the president created Operation Warp Speed, a small working group whose inflexible mission was to have a vaccine that could be distributed to the American public by January 2021. It really was Mission Impossible.

In addition to Secretary Azar and some administration appointees, Deborah, Tony, NIH Director Francis Collins, and I served as members of the board of Operation Warp Speed, the group formed by the president to accomplish this mission. Our task was to look at all the candidates in development anywhere in the world and identify those most likely to be successful so the government could direct its full effort—and billions of dollars—to facilitate their development, testing, approval and distribution.

We considered one hundred twenty different vaccines from laboratories around the world. These included candidates from China and Russia. We weren't focusing on a US-made vaccine. We wanted a vaccine we could deliver to the American public in months. Every one of them had the same target, the so-called "spike protein" in the Covid virus. If you could deliver the spike protein into the cell structure to produce antibodies, no matter how you did it, you won.

We began by cutting those vaccines that could not be produced—even if they were efficacious—and made available to the public by the beginning of 2021. That included those drugs that weren't far enough along in the process as well as companies that didn't have the necessary resources to move that quickly. The question we asked was: If we provided sufficient funding, could you manufacture 100 million doses within the next six months? That question allowed us to eliminate almost 90 percent of the candidates. Eventually we got down to about twelve that could be available—if they worked—within months.

Committing funding came with risks. There was no guarantee that any of these vaccines would be effective. To include the widest range of technologies, we decided to segregate the candidates by platform. That eliminated some duplication. For example, we looked at the companies working with mRNA and debated which of them could be ready by 2021. Of that group, we decided to fund vaccines being developed by Pfizer-BioNTeach and Moderna.

Then we looked at the vector vaccines. Vector vaccines take a small piece of protein from the COVID-19 virus and insert it into a modified version of a different, harmless virus. That virus, the vector, smuggles the protein into the nucleus and instructs the cell to make copies of it, which produces antigens to alert the immune system to the presence of a new pathogen. Tony Fauci and I had worked with Johnson & Johnson to develop a vector vaccine against Ebola and actually proved, working in a war zone, that it worked. We picked J&J and AstraZeneca from this platform.

The third platform was the traditional protein-based vaccine. Both mRNA and vector vaccines work by provoking your body to produce the antigen. The old-fashioned way that we have been doing for half a century is making the protein with an antigen and injecting it into your arm. Basically, researchers have to figure out how to grow the virus. They begin by getting the active pharmaceutical ingredients they will need. Most of them come from India or China. The virus actually is grown in eggs. Once a sufficient amount is produced, they have to determine which part of the virus should go into the vaccine. Then that segment of proteins would have to be formulated and tested to see if it spurred production of the desired antibodies and was safe and nontoxic. Assuming that was successful, they then had to determine how to get it into the bloodstream at the necessary levels. Producing a new flu vaccine every year takes about nine months.

The beauty of an mRNA vaccine is that it can be synthesized in a laboratory and injected. Boom. Done. I don't have to figure out how to formulate the levels your body needs to fight the infection, I just give you the mRNA and your body will make the vaccine.

Protein vaccines are not sexy, but they have distinct advantages. They are inexpensive to produce and don't have to be refrigerated. I was a strong advocate for protein vaccines because the basic science had been working successfully for more than a century, so we knew it was dependable—even if it took a bit longer. We also knew that these vaccines could rapidly be circulated around the globe. If we were going to contain Covid, we needed a global immunization strategy, and I believed this category made the most sense. The leading companies were Novavax and Sanofi Pasteur-GSK.

After considerable debate, to my disappointment, the mRNA and vector companies were selected for accelerated government funding. Speed mattered, and creating protein vaccines took the longest amount of time. But within a few months, we added the other two.

Then we waited, and watched, and looked at the initial trials, and answered questions, and removed potential roadblocks—all while maintaining pretty strict scientific guidelines. By early summer, stage one clinical trials were beginning. A few volunteers were getting the vaccines.

Our objective was to prevent as many deaths as possible while this was going on. Shutting down the country to reduce infection was a necessary first step. That shutdown would be extended several times. But as soon as it began, we started searching for safe ways to reopen the country. Our first recommendations, long before the country began shutting down, were the most obvious and simplest ones: Wash your hands and stay home if you don't feel well.

There is no better weapon against infectious diseases than clean hands. Most pathogens get transmitted hand-to-mouth. So the best way to protect yourself is standard sanitation. It's the same message public health agencies have been advocating for one hundred years: Wash your hands and wash your hands.

Unfortunately, in the initial weeks of the pandemic that concept got magnified to include cleaning everything that came into your house, including mail, packages, and groceries. There was some reason people were overly cautious; we still knew very little about this particular virus, and we did have evidence that other pathogens could live on surfaces—doorknobs, for example—and be spread by contacting them then touching your eyes or mouth. I remember various TV "medical" experts suggesting people wash everything that came into their homes just to be safe. Joy and I did not. I just hadn't seen any evidence Covid could be transferred that way. It wasn't a terrible idea. If this caution did any good at all, it made people more aware that exaggerated hygiene—washing your hands, wiping your boxes of new wipes—at that point was the best suggestion we had for people trying to protect themselves.

Our second recommendation was equally obvious: If you don't feel well, stay home. Don't go to work. That seems obvious, but historically that just hadn't been the American work ethic. A lot of Type A personalities felt they were tough enough to go to work even when they didn't feel well. They would suffer through the day at the factory, the store, or the office for the good of the company. That was precisely the worst thing a person could do. Even before we had a test, we were advising people with any symptoms to stay home in, if possible, a well-ventilated area. Get plenty of rest, stay hydrated, and stay away from everybody else. Isolate yourself if at all possible.

That sensible advice may have done some good, especially in those first few weeks. But it became obvious after the initial shutdown that any strategy allowing businesses to reopen necessitated the use of face masks. That realization was followed almost instantly by the recognition that a large number of people were going to object to having to wear a face mask.

The history of wearing masks to prevent infection is spotty. There is some evidence that during an outbreak of the plague in the early 1600s physicians in Paris dressed themselves with a leather costume that included a birdlike face mask with a beak filled with pleasant fragrances. King Louis XIII's physician described it as having a "nose half a foot long, shaped like a beak, filled with perfume with only two holes, one on each side near the nostrils, but that can suffice to breathe and to carry along with the air one breathes the impression of the drugs enclosed further along in the beak."

While those leather masks offered at least some protection, there is nothing that indicates Parisians knew what they were protecting themselves from. The concept of wearing a mask for protection from . . . something eventually compelled Louis Pasteur to agree to visit Egypt to study an ongoing epidemic—but only if, in addition to other precautions, he could cover his mouth and nose with a gauze-cotton mask. That response was mostly met with ridicule. During an epidemic, it was pointed out by leading physicians, "such precautions would be impracticable or illusory."

It wasn't until 1898, when research began indicating that pathogens could be transmitted by droplets from the respiratory system, that doctors even began wearing masks to reduce the possibility of spreading germs. The concept came from German hygienists, who suggested that while operating doctors should wear a "mouth bandage," a single layer of gauze that might capture potentially dangerous droplets.

At first that concept was ridiculed. As one surgeon wrote, "Due to our experience of many years we consider mouth masks—by the way, quite irritating—altogether unnecessary." Not everyone agreed. In 1905, Dr. Alice Hamilton, who pioneered the study of diseases being linked to certain occupations, and who later became Harvard Medical School's first female professor, saw that healthcare workers wearing face coverings while treating scarlet fever patients did not get sick and began urging physicians to wear masks in surgery.

The first widespread use of masks specifically to prevent infection took place in 1910. When the pneumonic plague—a disease that is almost 100 percent fatal—broke out in Manchuria that year, British-educated Chinese physician Wu Lien-teh speculated it was an airborne disease spread from person to person, rather than by the rodents that had caused previous epidemics. To combat it, he designed head-to-toe body coverings for his North Manchurian Plague Prevention Service inspectors. These uniforms include padded cotton and gauze masks that could be tied around the wearer's head with extra strings to "prevent it from slipping down the neck." They proved effective. Not a single member of the inspection teams contracted the disease. Subsequently, nonmedical personnel began wearing them. A factory in Harbin produced 60,000 masks, which were given away.

The idea spread, and gauze masks became common throughout the world during the Spanish flu epidemic. In the 1920s, doctors in Germany and the United States began wearing surgical masks in operating rooms. It took two more decades—and another world war—before the medical profession adopted the use of washable, sterilized masks. The kind of single-use disposable masks made of synthetic materials that we had available when Covid struck first came into general use in the 1960s. The problem was we didn't have enough of them to supply our frontline workers who were at a high risk of exposure. The front line included doctors, nurses, healthcare workers, and hospital employees. Moreover, those masks almost exclusively were being manufactured in China, in factories that had been shuttered because workers were ordered to stay in their homes. And China needed those masks that were being produced to protect its own people.

We also recommended from the very beginning that infected individuals—by our definition people who were symptomatic—be tested, isolated, contact tested, and wear masks until their symptoms were gone.

As I remember, there wasn't a lot of discussion about masks at the beginning. Masks have always made a great deal of sense to me as a valuable means of protection against infectious diseases. It seemed obvious to me. The data we had actually confirmed that if a person who was infected and symptomatic wore a mask, that alone would be very effective in preventing that person from spreading the disease. It was logical: While the mask wouldn't stop every particle, it would greatly decrease the amount going into the air that could potentially infect others. But we had no evidence showing that wearing a mask would help non-infected people from becoming infected. Logically, it also made sense: Two masks meant two protective barriers. Together, they could further stop the spread of infectious particles. If I'm infected and wear a mask, I am protecting you. If I'm not infected and wear a mask, I am protecting myself. We knew this was true even before we had the data to demonstrate it.

No matter what I believed, the initial guidance about masks came in late February from US Surgeon General Dr. Jerome Adams, who tweeted, "Seriously people – STOP BUYING MASKS! They are NOT effective in preventing general public from catching #Coronvirus, but if healthcare providers can't get them to care for sick patients, it puts them and our communities at risk!"

I agreed with that recommendation. It made sense. We were desperately short of masks and needed to reserve whatever supply we had for healthcare workers. More importantly, there was a such a high number of asymptomatic infections that most people didn't know if they had the virus or not. There really was no reason for someone not working in healthcare or taking care of infected people to wear a mask.

The message we were trying to communicate was pretty clear: If you're infected or in possible contact with infected people wear a mask. Otherwise, there is no benefit, and we needed every available mask. This was totally consistent with the guidelines established by the World Health Organization.

On March 6, President Trump visited CDC headquarters in Atlanta to announce the government was committing $8.3 billion to fight the pandemic. During that visit, none of us—the president, myself, or Secretary Azar—wore masks. We had all been tested, and the results were negative. Unfortunately, the president did wear a red MAGA hat. I thought that sent the wrong message. We were trying very hard to avoid any form of political partisanship. We desperately needed a unified national effort, and the country was deeply divided politically.

While the president was in Atlanta, he also mistakenly told reporters that anyone who wanted to be tested could be tested. That wasn't accurate. We weren't there yet. It got corrected pretty quickly, but it was the beginning of the mixed messages that the government was sending.

My relationship with President Trump was overall very good. He was getting advice from a lot of different people. He called me a number of times. It was a big deal for me to get a call from the president.

The first time the phone rang and an operator told me to "please hold for the president," I had to sit down. He would call me when he had a question because he wanted my opinion. He never acted in a way that suggested he was telling me what he wanted my opinion to be.

I wouldn't have done that anyway. I remember telling him, "My job, Mr. President, is not to tell you what you want to hear. My job is to tell you what I think you need to hear." When I was about to tell him something that I knew he wasn't going to like, I would begin by repeating that. We had conversations, not arguments. There were times he challenged me, telling me other advisors did not agree with me. And I would respond by providing the source of my opinion: the data or the science.

These weren't easy conversations. They were business. They were tough calls with serious questions. Sometimes he suggested I speak with people who had a different perspective, which I would do. But my interactions with the president were always civil and focused.

We certainly had disagreements. There were times he did not like my explanation, but he never raised his voice or challenged my opinion. He did do that on occasion in public comments, but never personally. I remember a discussion we had about mortality rates. There was a debate in the media about whether these people were dying from Covid or from other factors complicated by Covid. That was a really controversial question and had political consequences. He wanted me to explain the numbers to him. "Robert," he asked, "are they really dying from Covid or are they dying because of a heart attack or something else?"

I explained, "Mr. President, the Covid caused their underlying heart condition to cause their death, but it's Covid that ultimately cost them their life."

Was I comfortable with the way the numbers were being reported?

"I am, yeah. The comorbidities aren't the reason they're dying. It's not asthma. They died with asthma that got exacerbated by Covid, so it's really a Covid death."

On another occasion, he asked me why we had a higher percentage of Americans dying per infection than countries like Korea or Sweden. "The reason our death rate is up," I said, "is not because our medical care is inferior. It's actually superior to many parts of the world. Our death rate is up because a significant percentage of the American public is unhealthy. We've got comorbidities. Thirty percent of the American public is overweight, and obesity is a big cause of mortality with Covid. Koreans aren't overweight."

We went back and forth on that. I think that was one answer he didn't like.

But perhaps more than anything else, we strongly disagreed over the importance of wearing masks. After learning in March that asymptomatic transmission was possible, we began modifying our recommendations. The task force debated it for several days. We knew it would add to the confusion, but the science had changed. We had evidence that a significant percentage of infections were caused by people showing no symptoms, and we knew that masks would have an impact on source control. We finally decided that everyone should assume they are infected and wear a mask. On April 3, the CDC issued new guidelines recommending that people wear cloth masks in public, especially in places where maintaining six-foot social distancing was difficult. Because we were still dealing with shortages, we published directions for making rudimentary masks at home.

This was when the real division started. President Trump told the media that his administration was recommending that Americans voluntarily wear cloth masks; but he added that, personally, "I don't think I'm going to be doing it. Somehow sitting in the Oval Office behind that beautiful Resolute desk and wearing a face mask as I greet presidents, prime ministers, dictators, kings, queens—I just don't see it for myself."

Almost immediately, it became a political issue. That complicated the situation. I don't think I appreciated how strongly people would object to wearing a mask. Not wearing a mask was viewed as support for President Trump. In June, the president told the *Wall Street Journal* some people were wearing masks because they disapproved of him. When Senator Joe Biden, the frontrunner for the Democratic nomination, began wearing a mask, the president attacked him for it, saying it looked "like a knapsack over his face."

I never advocated for a national mandate forcing everyone to wear masks in public. Even if I wanted to, I didn't believe the CDC had the regulatory power to require people to wear them. I am generally against mandates unless, as in the case of the cruise industry, the evidence exists that forcing people to do something would have a universally beneficial effect. I also know from history that Americans have never responded well to being told they had to do something. That started with the Boston Tea Party. Mask mandates didn't work especially well during the Spanish flu; people were arrested for refusing to cover their mouth. Songs were written with lyrics like, "Obey the laws, and wear the gauze. Protect your jaws from septic paws!" There was absolutely no reason to believe things would be different this time. Congress probably could have enacted a national mandate, but it was far too controversial for that level of regulation to pass.

States did have that power though. In early April, individual states began mandating the use of masks in public. By the end of July 31, states and the District of Columbia had passed regulations mandating face coverings. We also were beginning to see scientific data proving that it worked. But that didn't matter. The response was just as I expected it to be. People were defiant, claiming their constitutional rights were being violated. They felt this was an affront to their personal liberty! There were several cases brought against the government, resulting in courts ruling that states did have the right to pass these laws. But this patchwork of requirements that varied from state to state, sometimes even from town to town within states, confused people and damaged the effort to create a unified national response.

I believed masks were the most valuable tool we had to protect people until we had a vaccine. I campaigned hard to make that point. And even after that, masks would continue to save lives. As I told Congress, "This face mask is more guaranteed to protect me against Covid than when I take a Covid vaccine because the immunogenicity (the protection) may be 70 percent, and if I don't get an immune response, the vaccine's not going to protect me." Then I held up a mask, more for emphasis than dramatics, "this face mask will."

In fact, I said, if all Americans would wear a mask for six or eight weeks we could bring the epidemic under control. But instead of forcing people to wear masks, it should remain voluntary. We needed to mount a massive publicity campaign emphasizing all the benefits, both individually and for society, of wearing a mask.

That never happened. The battle of masks continued with growing intensity throughout the entire pandemic. Personally, I wore a mask when I was going to be in a closed space with other individuals and when I was travelling. I'd wear a mask in indoor places where it was difficult to maintain social distancing. I'd wear a mask as I walked through an airport or passed through security. I'd wear a mask in crowded rooms or generally when I was going to be in close contact with people I didn't know. There was an overreaction that didn't help; for example, I'd see drivers alone in a car wearing a mask. That made no sense. I'd see people walking outside in the fresh air wearing a mask. That made no sense. Overuse like that just provided ammunition for the anti-maskers.

Believe me, I understood the criticism. I understood the complaints. In combination with social distancing and the general shutdown of society, the masks created a sense of alienation and isolation. All the normal social cues we get from looking at someone's face disappeared under the mask. I felt that was especially damaging to children, who are learning how to communicate.

For many people not wearing a mask became a matter of personal pride. People would attend large rallies protesting government overreach without wearing masks; without question, this spread the virus.

The president sent the wrong message. I can't explain his reasoning. Maybe he thought he was protected because every person who met with him was tested before that meeting? Maybe it was a campaign decision to appeal to a bloc of voters? I don't know. But he would regularly attend media events without a mask. In May, he wore a mask with the presidential seal on it when he visited a Ford plant in Michigan. But before meeting the media, he took it off, explaining that he "didn't want to give the press the pleasure of seeing it."

There was nothing we could do to change his mind. He began holding large campaign rallies at which few people protected themselves. Masks were optional, and thousands of people stood close together. For someone so deeply concerned about public health, it was disheartening. One of his first rallies was held in mid-June in Tulsa, Oklahoma. Among those people attending it was Herman Cain, the former CEO of Godfather's Pizza who had been a Republican presidential contender. Cain had supported social distancing and wearing a mask as the best route to reopening the country. But days before the rally, he posted on Facebook, "Masks will not be mandatory for the event which will be attended by President Trump. PEOPLE ARE FED UP!"

Unfortunately, a week after attending the rally without masking Cain was diagnosed with Covid. His condition deteriorated, and he died. While his staff pointed out it was impossible to pinpoint where he was infected—and that his immune system had been compromised years earlier by his fight against colon cancer—at least eight Trump campaign staff members tested positive after the rally.

But the people surrounding the president generally did a good job keeping him away from people who were healthy but who had not recently been tested. At least they did that until the first days of October, when, after an outdoor White House ceremony to celebrate the nomination of a new Supreme Court Justice, seven people tested positive. Both President Trump and First Lady Melania Trump were diagnosed with Covid.

As soon as the president was diagnosed, his doctor, Sean Conley, called me to discuss how best to treat him. Obviously, I was very concerned because he was seventy-three years old and not at his ideal body weight, which put him in the high-risk group. I told Sean he needed to be hospitalized, where he could receive respiratory support if he needed it. I suggested he tell the president that Walter Reed had a really nice presidential suite; I knew that because when I was sixteen years old, I had a summer job as a janitor in that hospital, and one of my responsibilities was cleaning out that suite. I also suggested treating him with the antiviral drug remdesivir, which appeared to have at least some effect in reducing mortality, and monoclonal antibodies produced by Regeneron that stimulated the immune system. Those antibodies were derived from a patient who recovered from the virus and from a mouse that had been scientifically engineered to have a human immune system and produced vital proteins. It was still an experimental treatment, but the preliminary results were promising. It was the best we had to offer at the time. I'm sure Dr. Conley was also speaking to other people during this difficult period, gathering as much information as possible.

The president agreed to be hospitalized. Sean and I spoke every day. Joy and I prayed for the president every day. Several days later, on a Sunday, Sean told me the president felt considerably better and wanted to be discharged. Please don't let him out of the hospital, I pleaded. "Sean," I said, "you and I both know that when people get over the viral stage from pulmonary infections, they start getting a lot better—then two or three days later they go into the cytokine phase, and they deteriorate fast." I had seen people go from feeling great to dead within twenty-four hours. The fact that the president was feeling great was not necessarily a good thing. I continued, "You've got to keep him there until Thursday or Friday, until he's past any risk of a cytokine storm."

"That's going to be a rough discussion," he told me. I understood the politics, but I was pretty sure the president didn't realize he was still sick. I offered to help Sean convince the president not to leave Walter Reed. This was not a benign infection, and his life was still very much in jeopardy. On Monday, he told me glumly that the president insisted on returning to the White House. We talked about how to get hyper-prepared in case something went south.

Then I held my breath. This could have been the beginning of a sea change that transformed the country. There was some part of me that hoped this experience would be an epiphany for him, that it would allow the president to appreciate and support what we were trying to do. That maybe, just maybe, he would become a megaphone for our public health message and end this damaging civil dichotomy. I watched the coverage of his arrival at the White House. The helicopter landed. When he emerged, he was wearing a mask.

He was wearing a mask.

A few minutes later, he emerged on the balcony still wearing that mask. Maybe, just maybe. . . . Then he ripped off the mask and waved. I was very, very disappointed. It broke my heart. I closed my eyes and sighed. This wasn't about protecting the president; he already was infected. The White House later released a video he'd made, "As your leader I had to do that," he said. "I knew there's danger to it but I had to do it. . . . Nobody that's a leader would not do what I did. I know there's a risk, there's a danger. That's okay. And now I'm better, and maybe I'm immune? I don't know. But don't let it dominate your lives. Get out there, be careful."

It was a terrible message. I don't know if the president knew how lucky he was. He was so sick he was given oxygen in the hospital. I'm convinced those monoclonal antibodies helped him recover.

President Trump was consistently inconsistent in his comments. He arrived too late to his first debate with Joe Biden to be tested—three days before he was hospitalized—and ridiculed his opponent, saying, "I don't wear a mask like him. Every time you see him, he's got a mask. He could be speaking two hundred feet away from them and he shows up with the biggest mask I've ever seen." But in July, he told supporters, "Whether you like the mask or not, they have an impact." At a rally just before the election, at which masks were optional, he advised, "If you get close, wear a mask. . . . It's not controversial to me."

He also deserves credit for making difficult important decisions. Our first meeting with him was held in the Situation Room. There were ten people there. We laid out the facts for him. We have a respiratory pathogen that is new to the human species. It's not influenza. He asked some questions. It wasn't clear to me that at that time he understood the seriousness of this presentation. His initial reaction to all of it was, well, this is interesting. I don't think he felt a real urgency. Within a few weeks, his attitude had changed. When we suggested shutting down air traffic from China, he made the decision to order that almost immediately.

Not long afterward, considering the stakes, he made what might have been his most difficult decision. During a task force meeting in the Oval Office, I told him we needed to shut down air traffic to and from Europe. He was surprised. "Stop it?" he asked.

"Yes. Stop it all."

Steve Mnuchin, the secretary of the treasury, was in the room and jumped up and warned, "You're not stopping travel from Europe. It's not going to cause a recession, it's going to cause a depression, and we won't get out of it for the rest of your presidency." Mnuchin's subtext was pretty clear: You've got an election in November, and if this tanks the economy it could cost you the presidency.

The president turned back to me and asked if I really believed we needed to do this. I told him, "Absolutely, and I wish I had made this recommendation two weeks ago. We just didn't have the data then—but we do now."

He turned to Deborah Brix, who was sitting next to me and asked her if she agreed with me. "I do," she replied. He looked at Tony Fauci and asked once again. Fauci agreed. Then he told Mnuchin, "Steve, we can always rebuild the economy, but we can't rebuild the lives that are going to be lost, so we're going to shut it down."

The president also supported me when I shut down the cruise industry, which we all knew was going to have a devastating economic impact, especially on Florida which had strongly supported him. He also didn't hesitate to commit billions of dollars to Operation Warp Speed, enabling this country to produce at least two vaccines in record time.

And in mid-March, at the suggestion of the task force he invoked the Defense Production Act to increase the production of N95 respirators, face masks, and ventilators. As a result, literally billions of face masks of all types—cloth masks, disposable surgical masks, and N95 respirators—were produced and distributed. It actually would have been substantially more. There was a plan in place to have the postal service deliver a package of masks to every household. But someone in the administration managed to derail that. While we never found out officially why that plan was canceled, supposedly it was done to prevent Americans from panicking.

I understood the president was trying to find a balance between his reelection campaign and protecting American health. Every decision he made reverberated through the electorate. The direct access we had to the president declined significantly when Scott Atlas became the White House liaison to the task force. I think Atlas was far more focused on the politics of the pandemic than the science. It was incredibly frustrating. We would meet with Vice President Pence sometimes five or six times a week, but we never knew how much information was getting to the president through him.

In my opinion—opinion—Scott Atlas did not understand the science and gave the president terrible advice. He believed there was no reason to expand testing. That there was no reason to try to contain it. That all we needed to do was protect the elderly, and eventually it would end. It was the total opposite of what we were trying to communicate. He was very aggressive. In one task force meeting, I remember so well, he told me in a voice pitched way higher than it needed to be that he was the person representing the president's policy and that what I was saying was not consistent with that.

Pence responded to that, saying, "Well, I am the vice president."

All of this was terribly frustrating to most members of the task force. Every day brought an avalanche of reports; every day was filled with highs and desperate lows. The death count kept rising. Having spent decades believing I was preparing for this, it became apparent there was only a limited number of things that could be done to prepare for it. In retrospect, my biggest disappointment was that science did not take the lead. When people look back at World War II, for example, the nation came together to fight the enemy. That didn't happen in this battle. Instead, there were too many camps with different agendas competing for attention and power.

There are many lessons to be learned from this experience, but none of them are more important than this one: In a crisis, the nation's leadership absolutely has to speak with a single, clear, and unambiguous voice. Decisions have to be explained in a manner that people can understand and support. Unfortunately, we did not do that. Because of that, many people suffered. And too many of them died.

SEVEN

I was sitting at my desk on a gloomy day in early November when the world changed forever. I was leafing through a briefing packet Steve Hahn had sent over before our scheduled task force meeting. Included in it, as if it was just a little more normal information, was a note that the FDA was approving the EUA, the emergency use authorization, for Pfizer's vaccine candidate. In stage 3 trials, the vaccine had a 95 percent efficacy. I had to read that again: 95 percent efficacy.

I was stunned.

Stunned.

This was a watershed scientific moment, the fact that we could deliver RNA effectively into the human body in a way to direct that body to make the protein we want—and the protein we wanted was the vaccine.

I had been following Pfizer's progress as much as possible. I had seen their phase 1 data that indicated they were able to get immunogenicity, meaning they were able to trigger the immune system to make antibodies, so we knew the basic technology worked. Efficacy—the question of whether these antibodies prevented infection—was a very different thing. The clinical trial included

43,000 people. It proved the vaccine was 95 percent effective in preventing hospitalizations or death.

Ninety-five percent! That was astonishing. The FDA had made 50 percent the threshold for approval for Covid vaccines; for other vaccines we were happy with 70 or 80 percent. This was unbelievable.

I just sat there, maybe I took a deep breath; for the first time, I could imagine an end to these deaths caused by this pandemic. We had taken the offense.

Days later, Moderna announced its vaccine had a 94.5 percent efficacy. For me, these results proved that this technology would revolutionize how we would develop and deliver vaccines in the future. My father's curiosity seven decades earlier had been vindicated.

We had been waiting for this since the first reports in January. We had been marking time, trying a variety of potential treatments to save lives. We were in a desperate situation. By the end of April, more than 50,000 Americans had died of Covid. Science hadn't provided any answers—we still didn't have a cure for the common cold—so we applied common sense solutions. And we made mistakes, and we learned.

Among the first steps we took in treating severely ill patients was putting them on ventilators. It made sense. Covid attacked a patient's lungs and made breathing difficult. Ventilators are machines that push air into the lungs, like a mechanical bellows. But we immediately encountered a serious shortage of ventilators in hospitals, more evidence that we weren't sufficiently prepared. It created a terrible, terrible dilemma for physicians who had to decide which patients would be ventilated. In some cases, this was a life-or-death decision; although, as it turned out not the way we anticipated.

States began competing against each other to purchase the few available machines. The price skyrocketed, almost doubling to about $50,000. In Italy, volunteers were using 3D printers to produce desperately needed ventilator components. At the end of March, President Trump invoked the Defense Production Act, ordering General Motors to start making these machines. After all that turmoil, all that scrambling, the data told us an incredible story.

Ventilators didn't save lives. A majority of the people who went on ventilators didn't come off. We didn't know why. That said, because of the shortage most people who were put on ventilators were already critically ill, and many of them also had underlying conditions like hypertension, heart issues, obesity, and diabetes. Still, it was disappointing. Rather than relying on ventilators, we tried to avoid employing them until it was absolutely necessary.

At the same time, we were learning the importance of other treatments. Although we had no clinically proven therapeutics, it became obvious very quickly that once people in intensive care got into the lung inflammatory stage, strong steroids, in many cases, helped shut it down.

The CDC had a clinical working group that put out advisories that were updated in real time whenever we got new information. We had calls every week with clinicians throughout the country, Clin Calls we called them, during which we would present any new evidence we had. Sometimes we had as many as 2,500 physicians on those calls.

We tried to create a forum for the exchange of information—a place where new techniques, treatments, and ideas could be introduced, discussed, and tried to help improve patient outcomes. One of the techniques that emerged from these discussions was turning patients on their stomachs early on, which made breathing much easier. I'm not quite sure where the initial suggestion came from. I believe a nurse recognized that patients who laid on their stomachs did better, tried it on several of her patients, saw that it improved ventilation, and then passed that information along.

That simple act, observation, has always played a vitally important role in medicine. I learned that when I started working with HIV. No matter what we did, all of our advanced patients were dying at twenty-five or thirty years old. What we were doing wasn't working. All the training we had, all the tools at our disposal, none of it was making any difference. So we had to be open-minded enough to try some things differently. That's the what-if factor that has often led to significant advances in medical research.

That nurse's simple observation made sense. Because of the position of the lungs in the body, closer to your back than your front, lying on your belly reduces the stress on the lungs. Turning patients over became an option for clinicians and fairly rapidly became routine for patients with breathing issues.

We were willing to look at anything and everything. Suggestions, thoughts, and ideas came across my desk every day. I've never been dismissive of another person's observations; I've always left open the possibility that an observation other people might dismiss could be a clue to something. Among those observations were anecdotal claims that vitamin D and type O-negative blood seemed to have some positive effect against Covid.

When we got an intriguing suggestion, we looked at our internal data to see if we could confirm or dismiss it. Then we debated if it was important enough to set a study to determine if it provided any benefit. Vitamin D, for example, has always interested researchers. A vitamin D deficiency has long been associated with disease, although there is no evidence it is a cause or even a contributor. There was nothing in our studies to confirm that vitamin D levels had any significant effect on the disease, although a preliminary NIH study hinted that it might reduce severity and mortality.

The antigens in different blood types had also been shown to play an undefined role in certain diseases. This came from a report out of Europe. A follow-up report seemed to show that people with O-negative blood were less susceptible to Covid. It didn't mean they were immune, just less prone to infection. So we examined it to see if it might give us a clue to a countermeasure. There just was no compelling evidence to confirm this observation.

We continued looking at all the data, trying to make connections. At the CDC we tried extremely hard to focus our comments on the science. We issued information sheets about numerous potential therapies, but they were basic, bland, and informational: This is what the drug is. Here are the ongoing trials. Here's how it is being used in those trials. Here are the results for other trials, always emphasizing that there was no definitive evidence that this was efficacious.

One day my colleague Bob Gallo called to tell me he believed he had strong evidence that Albert Sabin's oral polio vaccine offered protection against Covid. In addition to being a friend and a colleague, Bob Gallo is an extraordinary scientist—so this certainly caught my interest. If it was true, it completely changed the trajectory. His hypothesis went all the way back to the Russian flu pandemic in the 1950s, when the parents of Dr. Konstantin Chumakov, who later became the associate director of the US Food and Drug Administration's Office of Vaccine Research and Review, conducted a study that showed people who had gotten the new polio vaccine were protected against influenza for at least a month. Dr. Chumakov actually was one of his parents' trial subjects. There were other examples of this phenomenon. In many cases, people who got the live measles vaccine also had protection against the flu. A seven-year Russian study including 320,000 subjects found that several vaccines reduced mortality from the flu. The hypothesis made sense: Covid was an RNA virus. Could another RNA virus—polio or measles, for instance—trigger the immune system to prevent the Covid virus from gaining a foothold? The protection only lasted for a limited time, so it was not a substitute for a Covid vaccine, but it might save lives while we were waiting for the dedicated vaccine.

It might also provide information for researchers. What was the cellular mechanism that prevents a second virus from taking hold in that setting? Answering that question could help enable us to identify the critical pathways cells used to prevent an RNA virus from successfully replicating in the cell.

I recommended strongly that we aggressively study this. Because of Covid, we had been forced to shut down some ongoing vaccination programs in sub-Saharan Africa, so it would have been relatively simple to conduct a controlled study. I spent months trying to convince groups to determine whether or not vaccinating people against polio or measles offered any protection against Covid, especially in areas of great poverty that would not get immediate access to a vaccine when it became available. The fact that by law Congress mostly funds projects designated by the CDC, rather than funding the agency itself, really hurt. If I had the freedom and the money to conduct independent projects this would have been at the top of my list.

I didn't, so I couldn't. I tried to get funding from the NIH, from my old group in the military but we couldn't secure the resources I needed. Instead, rather than running a controlled clinical study, the gold standard in scientific research, we set up an epidemiological study in which we collected and examined data from places in Africa where our vaccination programs were ongoing. By the time we had sufficient data to begin reaching conclusions, Covid vaccines had been created. But later studies did support Bob Gallo's hypothesis that viral interference, via the oral polio vaccine, did offer at least temporary protection from Covid in children.

This type of creative research was taking place in countless laboratories. By the beginning of summer, there were at least 950 clinical trials in progress; at least 450 were already actively recruiting patients. The world of science had been mobilized. As a scientist, it was incredibly exciting to be part of this massive effort to unravel the mysteries of a previously unknown virus. But as someone who had access to all that information, at times it created tremendous anxiety. That number, 2.2 million dead Americans, was always in the back of my mind.

We also were dealing with an American public that had been spoiled by the wonders of science. Computers had provided extraordinary access to information and instant communications that we just didn't have during the HIV epidemic. So much had happened so quickly—drugs had been discovered that fought cancer, grew hair, restored sexual performance, and even transformed AIDS into a chronic illness rather than a death sentence—that people anticipated this virus would be quickly stopped too.

There rarely are eureka moments in science. Discoveries, advances, solutions, and cures almost always are the result of a long chain of research. During both the SARS and MERS outbreaks, there had been a good deal of research to try to find a vaccine; but because both of those viruses had been limited and controlled, no vaccines had been produced. That said, some really good work had been done. So we looked to see if there was something on the shelf that might provide some benefit. The first drug we found was remdesivir.

Remdesivir had been developed by Gilead Sciences in collaboration with the CDC and the Army's Research Institute of Infectious Diseases to treat hepatitis. It had also been investigated as a viable treatment for people infected with RNA viruses like SARS, MERS, and Ebola. It turned out to be the first treatment for severe Covid. In March, the administration made it widely available for compassionate use—for people who had nothing to lose.

Its efficacy was questionable. Studies showed mixed results, at best. The WHO essentially reported it had little value, although Gilead dismissed those results because the study included results from different countries with varying healthcare standards. But we had anecdotal reports that in some cases it appeared to have some positive effect. And for a time, it was all we had.

In March, the FDA, perhaps showing our desperation, also permitted the emergency use of hydroxychloroquine, based on reports that a laboratory had shown it was effective against infected cells. It was not unusual for doctors like myself to use a drug approved for one purpose to treat a different disease, so it was perfectly fine to investigate it in small trials to see if it had any value. This was evidence we were willing to try anything that showed promise. Hydroxychloroquine was essentially one of the oldest effective treatments in history. In 1638, an Incan herbalist in Peru had used the powdered bark of a tree to heal the wife of a Spanish Viceroy who was suffering from malaria. It took almost two centuries before quinine, the active ingredient in that treatment, was isolated. To reduce its toxicity, it was put into a compound which became known as hydroxychloroquine, and it proved valuable in a variety of applications, including lupus and forms of arthritis. Like some other drugs, it isn't known exactly how it works, just that it works. This led to the laboratory experiments that produced results sufficiently encouraging for the FDA to give it approval for limited use.

The fact that it received approval sent a message to the increasingly anxious public: We're trying. We'll find something. We get it. President Trump began actively supporting it. I don't know why. But his supporters joined him, and very quickly believing that this drug was an answer became a political loyalty test. The state of Oklahoma, for example, a strong Republican state, spent $2 million stockpiling the drug.

There was a lot of excitement about it. In March, I appeared on Anderson Cooper's show on CNN to discuss it. Asked if I would recommend it, I responded that is a decision between the doctor and the patient, and like CNN's Dr. Sanjay Gupta, "We're very comfortable in responding when we have data that is compelling, and I think this is what the studies are going to give it. CDC . . . is not an opinion organization. We're a science-based, data-driven organization."

Anderson Cooper then wondered about the dangers in recommending it, comparing it to what happened in the early days of AIDS. "You worked," he said, "did a lot of really important work in HIV. Isn't there a danger with the drug like AZT? A lot of gay men tried AZT . . . and it ended up killing a lot of people."

I agreed that was a great analogy. "That decision was premature... This is why . . . from a scientific point of view, from a data-driven point of view. . . . The way to do this is to do the appropriate trials and get the answer."

That wasn't good enough for the growing opposition. The day after that interview, Fox's Laura Ingraham attacked me, telling her viewers I had been on CNN "essentially dismissing, trashing, a backhanded slap at hydroxychloroquine despite all of its success stories. Unbelievable." Then she suggested I should be fired.

The next morning Secretary Azar called me. "Bob," he said, "I had nothing to do with that."

"Why would I think that?" I asked.

"Well, you know," he admitted, "I'm really good friends with Laura Ingraham. I thought you might think that I sort of set that story up."

Truthfully, that hadn't occurred to me until he said it. But when he made a point of it, it became impossible not to wonder about it.

The clinical trials continued—the results were not good. Hydroxychloroquine did not prevent infection, it wasn't an effective treatment, and, worse, it appeared to cause heart problems in some Covid patients. Within months, the FDA had revoked its emergency use approval. Unfortunately, in his own haste to reassure and calm Americans, President Trump began touting hydroxychloroquine in his almost daily appearances with the task force. That should not have happened; the president should not have become a cheerleader for this drug until we had some evidence, any evidence, it had medical value. But rather than listening to experts who told him there was little evidence it had any effect, he listened to his economic advisor, Peter Navarro, and began touting it. "It's a very strong, powerful medicine, but it doesn't kill people," he told Americans. "We have some very good results and some very good tests. . . . But we don't have time to go and say, gee, let's take a couple of years and test it out. And let's go and test with the test tubes and the laboratories. We don't have time. I'd love to do that." It was incredibly damaging to our continuing effort to present clear, concise scientific information to the public. For whatever reasons, the president championed this drug, essentially ignoring the fact that for this virus it was worthless. Worse than worthless. While the numbers remain hazy, there are claims that thousands of people died from taking this drug.

As we were to witness, it got significantly worse several weeks later. At the task force meeting on April 23, the president was told about research showing how light and humidity affected the spread of the virus. Later that day, he participated in a press conference. The president was front and center at these press briefings almost every day. One reason he did it was to refute claims that he wasn't taking Covid seriously. By that point, I was no longer involved in any public White House press meetings. I was watching on TV as Bill Bryan, the head of science and technology in Homeland Security, explained how temperature affected the virus and the benefits of being outside. He explained that, "As a scientific community, we are continuing to study the virus to understand its characteristics. . . . We've identified some of the weak links in that chain that the transmission of the virus depends upon. We identified that heat and humidity is a weakness in that chain. We've identified that sunlight, solar light, UV rays is a weakness in that chain. . . . Science is a process. . . . It is what it is. But we are now starting to get results and every week or two weeks, we're starting to find out something new and something different." Then he outlined some studies examining the effect of disinfectants on the virus. "We tested bleach," he said. "I can tell you that bleach will kill the virus in five minutes." That wasn't unusual, people were still using bleach to clean objects brought into their environment.

The president didn't seem to understand that. He told the media, "A question that probably some of you are thinking of if you're totally into that world. . . . So, supposing we hit the body with a tremendous—whether it's ultraviolet or just very powerful light—and I think you (Bryan) said that that hasn't been checked, but you're going to test it. And then I said, supposing you brought the light inside the body, which you can do either through the skin or in some other way, and I think you said you're going to test that, too. It sounds interesting. And then I see the disinfectant, where it knocks it out in a minute. One minute. And is there a way we can do something like that, by injection inside or almost a cleaning? Because you see it gets in the lungs, and it does a tremendous number on the lungs. So it would be interesting to check that."

I sat there in disbelief.

The president of the United States of America had just wondered aloud if people should ingest poison. It was a terrible moment. He turned to Deborah Brix and asked, "Have you ever heard of that? The heat and light relative to certain viruses, yes, but relative to this virus?"

She was being put in an awful position. "Not as a treatment," she said. "I mean, certainly fever is a good thing. When you have a fever, it helps your body respond. But I've not seen heat or light as a . . ."

I suspect she was relieved when the president interrupted her, suggesting, "I think that's a great thing to look at. Okay?"

But the damage was done. I like to believe if I had been sitting there, I would have responded more forcefully. Maybe that's why I wasn't there. But I like to believe that I would have said, "Mr. President, you didn't mean to recommend that. There's no indication to do that. In fact, it would be dangerous." Personally, I believe the president was joking, but unfortunately many people didn't take it as a joke.

The president's comments were bad enough, but they did get exaggerated: The president recommended you drink Clorox to kill the virus. He never said that, but that's what the internet heard. There is no question that some people followed that advice, although the precise number is impossible to determine. But there was a significant rise in the number of people treated for ingesting common household disinfectants and bleach. There were reports of deaths.

The next best drug candidate, ivermectin, proved to be hugely successful against Covid—on the internet. In actual use, it was much less beneficial. The internet has innumerable benefits, but it also is the best tool ever created for snake oil salespeople. There are no filters, no easy ways of determining the validity of a post. Users can make outrageous claims, and there will be some people who believe them. Ivermectin actually is a safe and wonderful drug. It was discovered in a soil sample found near a Tokyo golf course and basically prevents parasitic larvae from growing. Through a program created by Merck, in which it literally gives away the drug to impoverished people in many countries, ivermectin has helped eradicate the terrible disease of river blindness. It also has proven useful in treating head lice and rosacea. It also is an extremely valuable veterinary drug. It is widely used in horses, cattle, sheep, and pigs in higher doses and different formulations to kill intestinal worms, mites, lice, and grubs. It is also used in dogs and cats to prevent heartworm. The discoverer won a Nobel Prize for it, and the WHO considers it an "essential medicine."

So people got excited when an Australian lab reported that in a culture of African monkey kidney cells, ivermectin had successfully inhibited the virus; within forty-eight hours, it had killed 99.8 percent of the Covid RNA. That was startling news! The drug had prevented the coronavirus from replicating. No one knew how it worked. But the results were extremely promising.

To play off a popular Las Vegas promotional line: What happens in vitro often stays in vitro.

The excitement increased a month later when a non-peer-reviewed preliminary study published online reported that the drug appeared to reduce mortality in human beings.

A few flakes of snow had become an avalanche, even as it was becoming clear in a variety of tests and trials that ivermectin had no effect on Covid. The Australian lab, it turned out, had used a massive amount of the drug to kill the virus, as much as ranchers used to treat 1,500-pound animals and more than enough to be toxic in human beings. In fact, the FDA actually warned people on its website, "You are not a horse. You are not a cow."

The preliminary study also turned out to be invalid. The data cited apparently was fraudulent. The study named hospitals that supposedly had conducted tests, but those hospitals denied any participation. Patients who had died before the study began were included. The same data was reposted several times. It was obvious this study had no value.

None of that seemed to matter. Where the science ends, politics and desperation begin. Social media exploded with all types of wondrous reports—none of which could be verified. Countries around the world embraced it. Fringe elements that had previously supported conspiracies embraced it, claiming the government was intentionally keeping it away from people for a variety of nefarious reasons. Popular podcaster Joe Rogan created a sensation when he told his 11 million listeners that ivermectin had cured his own case of Covid. Ivermectin was readily available over the counter in livestock supply stores; in some places, it sold out. The fact that no studies found it had any value made no difference. An entire pro-ivermectin industry, largely made possible by the internet and media, took root and grew—an early and ominous indication of what we were to face with vaccines when they finally became available.

Among other existing drugs that people suggested might have some value was, believe it or not, Viagra. It actually made some sense. Viagra is a drug known as sildenafil. In addition to erectile dysfunction, it is prescribed to treat pulmonary arterial hypertension. That is a complication of acute respiratory distress syndrome, a life-threatening condition common to seriously ill Covid patients. But trials showed little benefit.

Large doses of vitamin C were tested. An Egyptian trial involving one thousand participants tested natural honey, based on the claim that natural honey was used as a treatment for coughing resulting from an upper respiratory infection. In India and Pakistan, people experimented with homeopathic remedies like Arsenicum Album 30c. In other countries, mixtures of medicinal herbs, including wormwood, sagebrush, tarragon, and neem leaves, were offered as possible treatments. Residents of Nepal read that turmeric and garlic could ward off Covid. Iranians resorted to camel's urine or olive oil. In several countries, desperate people tried to protect themselves with aspirin, an analgesic useless against a virus, mixed with teas and juices. It was suggested that various fruits and spices, seasonings, and even chili could prove beneficial. Obviously, none of them proved to have any value.

And naturally there were legions of scam artists offering quack cures. One far-right celebrity podcaster sold "nano-silver" toothpaste. Other people were selling herbal remedies, vitamins, and supplements. Another media personality was selling oleandrin, a completely unproven extract from the oleander plant, which he claimed HUD Secretary Ben Carson had told him was "the real deal," had been tested on one thousand people, and had passed peer-reviewed public studies. A German beer company advertised a beer that worked as an oral vaccine. A South Korean company was selling a car air purifier that supposedly killed 99 percent of bacteria, including Covid—even though Covid was caused by a virus rather than bacteria. A Chinese company was offering special necklaces and static-free clothing which purported to offer protection. A Utah resident eventually went to prison for three years for scamming desperate people out of more than $6 million, selling them silver-based treatments for arthritis, diabetes, pneumonia, the flu, and Covid-19.

It seemed like years had passed, rather than months, since those early days when we believed—hoped—we might be able to control the spread of the disease. We were in the middle of a pandemic. Large parts of the country—including schools, factories and offices—had been shut down, and every decision was tainted with politics. So many of those decisions were made by governors who embraced and adopted policies that were politically expedient for their own ambitions. What that meant was we ended up with a patchwork of regulations that differed, sometimes wildly, from state to state. Once again, when a unified national response would have proved beneficial, we ended up in a disjointed and confusing situation. The fact that we didn't have the data we needed to provide simple straightforward advice created the void that allowed so many different—but sometimes less accurate—voices to be heard.

My own relationship with Secretary Azar continued to deteriorate. It had begun poorly when he publicly questioned my salary, which he personally had approved, and it had never really healed. The first real confrontation we had was in late February, when I was informed by people in Seattle that the first American had died of Covid. I had been up almost the entire night speaking with people all over the country who were reporting case after case after case. That was new. Until that moment we had been dealing with it almost by individuals. One case in one place. That changed in one day. Suddenly I was being told about multiple cases in multiple cities. And it was late into that night that I realized this epidemic was out of control.

At our task force meeting the next morning I went through all the cases with the president and vice president. "We've had the first Covid death," I said. I identified that first victim as a woman. During his televised press conference, the president then identified the victim as a woman, "a medically high-risk patient in her mid-50s."

The victim actually was a man. The president had to be corrected. I immediately took responsibility, tweeting that the CDC had "erroneously identified the patient as a female." It was a mistake. Maybe my mistake, but I certainly took responsibility for it.

Azar jumped all over me. He was furious, far beyond what was necessary to make his point. It was my fault. I got crucified for misleading the president. It mushroomed into a major issue in the media—and in the White House.

But not to the president. I don't remember him ever mentioning it to me.

It was my belief that the politics of Covid were more important to Secretary Azar than the science. I didn't care about not being allowed to be interviewed on national TV. Big deal. As long as we were getting an accurate message out. But our biggest arguments came over content published by the *Morbidity and Mortality Weekly Report*. The *MMWR* is the CDC's most important publication. It is circulated within the entire medical community. We describe it as a "scientific publication of timely, reliable, authoritative, accurate, objective, and useful public health information and recommendations."

But there are rules. It is an independent publication. It is run by an editor and an independent board. Even though it is published by the CDC, the director does not have the right or power to dictate to the publication board. It has been run like that since its founding. On more than one occasion, Secretary Azar demanded that I interfere with that publication. For example, it ran a story highlighting the severity of the outbreak that was not in line with the administration's optimistic message. He wanted it to be written in a manner that was more reassuring. It wasn't a matter of right or wrong, or even nuance. They wanted me to become, in more ways than one, a censor. I knew that if I went down that path, it could too easily be turned into a superhighway. What assured the publication's continued respect by the community was its editorial independence. That wasn't going to change, at least while I was director.

This actually was a very unusual approach. All the previous outbreaks the CDC had dealt with, even potentially serious ones like Ebola, remained within the CDC. This time was different. At the end of January, Secretary Azar had declared that this was a national public health emergency and assumed a central role. The Food and Drug Administration, the NIH, the CDC, and even the Surgeon General were all underneath his leadership. The White House appointed him head of the task force, although within a few weeks Vice President Pence took control.

For me, that meant the guidelines we issued had to be circulated to Azar's office for input. Input—not approval. I had done that, but I was not going to let him dictate changes. When I refused, his lawyers called me explaining why I had to do what the secretary demanded. It got very contentious. My job was threatened. I replied that the only way they could get the director to change the MMWR would be to get another director.

The most difficult confrontation we had concerned a recommendation from the Advisory Committee on Immunization Practices. ACIP had been formed more than half a century earlier to advise the government on vaccine policy. It is a highly respected group of healthcare professionals who are experts on this subject. Basically, the committee makes policy recommendations. after those recommendation are approved by the director of the CDC, they are published in MMWR the next day. It is the advisory for all the local health departments. While the vaccines were still in development, ACIP began planning the distribution schedule: which groups got it first, second, and so on. That decision was based on computer models that examined various scenarios and projected the number of lives that could be saved. ACIP held dozens of meetings with experts in all areas of vaccine production and distribution. The committee recommend that healthcare workers should get it first, followed by residents of long-term care facilities.

I approved it. When I informed Azar, he got really upset that I hadn't asked for his input before making that decision. I reminded him that my responsibility was to inform him, not to solicit his advice. After that, things got complicated. As we got closer to actually rolling out the vaccination program, ACIP continued to recommend this approach, and I continued to approve it. My only obligation to Azar was to inform him of ACIP's recommendation. His feedback made no difference. ACIP has always been an independent committee.

The secretary did not agree with it. He advocated immunizing the elderly and most vulnerable people first. He essentially told me that I should not accept ACIP's recommendation, that I should change it and publish that position in ACIP.

There is no simple right or wrong about setting a vaccination schedule, especially at first when there is a limited amount of the vaccine. It literally is a life-or-death decision. This recommendation made sense to me. Initially, I anticipated having somewhere between 4 to 10 million doses. The secretary said we were going to have 100 million doses. ACIP's number one goal was to maintain the resiliency of the healthcare system. Healthcare workers were the most vulnerable to infection. And after they were exposed, they had to go home. In addition to potentially spreading the disease, they couldn't work to save others if they got infected. These were essential workers. These were the men and women maintaining the entire healthcare system. If we lost them, we lost the ability to treat and protect patients.

Azar disagreed. I suspect there were people in the administration telling him they wanted to prioritize elderly, vulnerable people. Protecting that population might cut down the number of deaths in the short term, but it would prolong the pandemic and lead to more deaths. I understood that. In my statement supporting ACIP's recommendations I added that "future recommendations, based on vaccine availability, demonstrate that we as a Nation also prioritize the elderly (>70 yo) who reside in multi-generation households."

That wasn't good enough. Secretary Azar wanted what he wanted. He called me into his office. Several other members of his staff were already there. In front of them, he began berating me, loudly, accusing me of being the first CDC director who was going to be responsible for the deaths of more than ten thousand people. It was a preposterous accusation. Absolutely preposterous. I was furious.

When I was younger, I had a quick temper. I'd risked my military career confronting my senior officer over AIDS policy, for example, While I had learned to control it, the anger still swelled inside of me. I wasn't about to be bullied. I would have loved to have stood up, told him off, and resigned, which might have been his goal that day, but I didn't think America was better off if I did that. I was the obstacle standing in his way, and there was nothing he could do about it.

So I left. Within hours, his chief of staff and his attorney called and began the conversation by telling me, "I'm glad we have an agreement. You're going to write that document."

"What are you talking about?" I asked. "There's no agreement. I'm not going to write any such thing. My recommendation stands as it is."

Literally, for the next hour they just ripped me and ripped me. They told me it was my responsibility to listen to the secretary. It went on and on. "You're wrong," I told them. "My responsibility is to the healthcare system, not the secretary." The last thing I was going to do was change the recommendations of ACIP and publish it in MMWR.

I had no idea why they believed I had agreed to any changes. I hadn't. I didn't. I would not. By then it had become clear that we were representing very different constituencies. Instead of the unified response this crisis demanded, the country was splintered, listening only to those people that they agreed with politically.

In that divided state, we all lost.

EIGHT

From the moment we began closing down the country, we began trying to figure out how to safely reopen the country. Once again, there were no historical lessons to teach us how to do that. In 1918, for example, after shutting down a vast number of businesses, schools and social events for slightly more than a month, the country decided the threat was over and went back to work and school. The result was predictable, a second wave of infections and death. Those places that kept many of the regulations in place had the fewest deaths.

We were facing tremendous social, economic, and political pressure to reopen. Millions of Americans were frustrated and angry. An estimated 43 percent of businesses were shut down. Almost 30 percent of small business workers were laid off. Unemployment skyrocketed to 14.7 percent. Factories struggled to stay open. People were worried about their finances, they were terrified they would lose their homes, they were suffering through shortages of essential products, and they were cooped up and missing their friends. They refused to wait any longer, they wanted their normal lives back. The administration was exerting pressure too; the presidency was at stake. We had no choice. To restore the economy, we had to open up as quickly as possible. We couldn't wait for the vaccines.

Once again, it was a life-or-death question. We knew how dangerous it was. In countries where lockdowns had been lifted too soon, like China, South Korea, and Germany, there had been a sharp spike in infections and deaths. As far as I was concerned, in addition to maintaining social distancing, wearing masks, and washing your hands, it was essential to improve ventilation. The evidence was literally in front of us: Infections went down when people were outside. Infections increased when people were in closed rooms with little air exchange. We issued guidelines, urging people to open the windows, use fans, maintain clean filters, and, when possible, use high-efficiency particulate air filtration systems. The government granted billions of dollars to school districts to enhance the quality of air in classrooms, although most districts were too busy running virtual learning programs to really embrace that technology.

I was spending a lot of time meeting with local and state politicians and various essential industry leaders to figure out how to operate as safely as possible. I was preaching ventilation like a Sunday morning tent revival evangelist. The airlines were especially tricky. The country relied on efficient air travel for both commercial and personal transportation, but no one wanted to sit for hours in a cramped space breathing recirculated air or wearing a mask. The airlines were losing a fortune, and some smaller companies had already gone out of business. Planes that were still flying were always at least half empty.

Basically, planes had been designed so the flow of air moved through the entire cabin—either front to back, or back to front, and then out. But that meant every passenger was breathing the same air. I met with the CEOs. They redesigned the air exchange so that it went vertically from the ceiling to the floor and then out, which made air travel extremely safe.

Reopening the cruise industry was considerably more complicated. While not essential, it is an economic engine providing thousands of jobs and putting billions of dollars into the economy. I was told many times by both political and corporate leaders. Their implication was clear: I was personally responsible for destroying this entire industry. I got it. But because potentially thousands of people were together in a mostly closed space, sharing facilities, it was hard to make it safe. The companies created a panel of experts to help mitigate the crisis. My initial "no sail" order was extended four times, and during that period we met with this group several times. These experts wanted me to let them decide how to safely bring that business back to life. Also, the administration made it clear to me they did not want me to renew my "no sail" order when it expired in October.

I fought back. I pointed out to them that Covid was still spreading, that almost 346,000 new infections were reported the previous week. 215,000 Americans had already died, and there was no question this would add to that number. By that point, we had data that showed every infected person on the *Diamond Princess* had subsequently infected fifteen other passengers, more than four times the rate of infection in Wuhan. I told them that in good conscience I couldn't lift that order and if they did not support my recommendation, I would have to resign.

Suddenly, I came up with a different approach. I realized I could change the "no sail" order to a conditional sail order, meaning I was not blocking ships from sailing—provided they met certain conditions. I put together a panel of experienced public health experts who came up with more than sixty conditions a ship would have to meet before being certified to sail. Those conditions included more frequent testing of passengers and crew, limiting the duration of a cruise to a week, and enforcing social distancing standards in cafeterias, restaurants, entertainment venues, and even poolside lounges, which had to be six feet apart. All self-service options, buffets, drink stations, and salad bars were eliminated. It would take cruise lines months to meet those conditions, which was our objective.

It worked out as I had hoped. For the most part, the industry remained dormant until we had vaccines.

There were people inside the administration who didn't want to wait for a vaccine. They felt we were being far too cautious. Instead, they advocated protecting those people at risk while completely opening up the country. They believed the result would be a sort of natural protection known as "herd immunity." Herd immunity means simply that when enough people in a group have gained immunity from an infection, whether that comes from being exposed to the disease and developing antibodies or being vaccinated, the virus can't find sufficient hosts and disappears. It is a proven strategy and works for certain diseases. Generally, it requires an extremely high percentage of people in that group to be immune. To achieve herd immunity for measles, for example, more than 90 percent of the group has to have developed protection.

Several people cited the Swedish example as evidence this would work. Sweden had embraced a passive response, banning large gatherings and encouraging people to practice social distancing and to take other sensible measures. Its restaurants, bars, shops, and schools remained open. The government did not impose mask regulations. The result was predictable: Sweden had a higher percentage of Covid deaths than its Scandinavian neighbors. It eventually imposed the same type of restrictions as most of the world.

Among the leading proponents of this strategy in the administration was an HHS senior advisor named Paul Alexander. "So the bottom line," he wrote in a July email, "is if it is more infectiousness now, the issue is who cares? If it is causing more cases in young, my word is who cares . . . as long as we make sensible decisions, and protect the elderly and nursing homes, we must go on with life."

In a different email, he suggested people expose themselves to Covid to help us create herd immunity, writing, "There is no other way, we need to establish herd, and it only comes about [by] allowing the non-risk groups [to] expose themselves to the virus. PERIOD."

He lobbied me too, telling me, "We essentially took off the battlefield the most potent weapon we had . . . younger healthy people, children, teens, young people who we needed to . . . infect themselves, spread it around, develop immunity, and help stop the spread."

Inside the White House, Scott Atlas was the strongest supporter of this concept, although he would later deny it. I don't think he fully understood what we were facing. In early March, months before he officially became a special advisor to the president, he sent an email stating that the government's response was "a massive overreaction" that was "inciting irrational fear" in the country. Although he explained that "the panic needs to be stopped," he estimated the death toll from Covid would only be "about 10,000."

Obviously, this was not the first time I disagreed with Atlas, although until late September I had managed to keep my opinion to myself. While I was flying from Atlanta to Washington, someone overheard a private phone conversation I was having with a colleague. It was my fault; I should have been more careful. NBC reported accurately that I said, "Everything he says is false." I certainly believed that. I thought he was feeding very misleading information to the president, and the benefits of herd immunity was one of them.

Atlas had brought "three distinguished infectious disease experts" to a meeting with Secretary Azar in the White House that was widely covered by invited media. This group supposedly represented almost six thousand "medical and public health scientists" who recommended reopening the country. Their Great Barrington Declaration stated that "Current lockdown policies are producing devastating effects on short and long-term public health. . . . Keeping these measures in place until a vaccine is available will cause irreparable damage, with the underprivileged disproportionately harmed." Open up, they advocated. Open businesses, and offices, and stadiums, and movie theaters, and restaurants, and schools, "while society as a whole enjoys the protection conferred upon the vulnerable by those who have built up herd immunity."

As Azar posted on Twitter, "We heard strong reinforcement of the Trump Administration's strategy of aggressively protecting the vulnerable while opening schools and the workplace." Not surprisingly, other people in the administration denied that was a strategy.

Herd immunity can work. It has been proven to be an extremely beneficial too—in specific and limited circumstances. But this was not one of them. This concept, that basically proposed that the sooner more people were infected with Covid the sooner the pandemic would end, was nonsense. We already had strong evidence this virus didn't work that way. For COVID-19, there was no possibility of herd immunity. The data made it clear that there was no durable immunity to this virus. By the beginning of May, I had been seeing patients who were having a second episode of the disease, so it was obvious to me that—unlike measles or polio—the immune system's protection was only temporary. That natural immunity to the initial strain of this virus lasted only a few weeks or months. And very early on, we were seeing variants that would require a different vaccine. So herd immunity was not a realizable goal.

But that didn't stop Scott Atlas from convincing a lot of people that we could get to herd immunity with 30 or 50 percent immunity. That was absurd. Herd immunity only can work when the vast majority of people in a group or area, usually 90 percent or more, have that immunity at the same time. Atlas, unfortunately, claimed a considerable lower percentage was all that was needed, causing Michael Osterholm, the director of the University of Minnesota's

Infectious Disease Research program, to marvel. "That 20 percent is the most amazing combination of pixie dust and pseudoscience I've ever seen." When it is transient—meaning by the time you get it, my protection is gone—it doesn't work. That was the reality of Covid. Even Tony Fauci, who had been suggesting that herd immunity might potentially be a viable strategy—once an effective vaccine was available and we could inoculate vast numbers of people within a short period of time—disagreed with that Declaration because it assumed it was possible to identify and protect people who might be vulnerable. "That doesn't work," he said. "By the time you get to herd immunity, you will have killed a lot of people that would've been avoidable." As he put it, "If you let infections rip as it were and say, 'Let everybody get infected that's going to be able to get infected and then we'll have herd immunity.' Quite frankly that is nonsense, and anybody who knows anything about epidemiology will tell you that that is nonsense and very dangerous."

It was worse than that. Tedros Ghebreyesus, the director-general of WHO, said letting the virus spread to try to create herd immunity is "scientifically and ethically problematic. . . . Never in the history of public health has herd immunity been used as a strategy for responding to an outbreak, let alone a pandemic."

I believed promoting herd immunity was dangerous, even if it did have some obvious political and economic benefits. But it essentially provided an excuse for those protestors who refused to maintain social distance, wear a mask, or take other sensible precautions. It allowed them to claim they were following the recommendations of experts and that they were supporting the president—by doing nothing to prevent the still-raging spread of the disease. I didn't know what the consequences of embracing this concept would be, other than the fact that more people would get infected, more people would die, and it would contribute nothing toward ending the pandemic. A talking head with a medical background claimed on CNN that an additional 2 million Americans could die if this strategy was adopted. Maybe. There was no way of predicting.

What was personally disappointing to me was that Tony Fauci held it out as a possibility. By the summer he had become the public face of the task force and appeared on TV every day to announce, or explain, or interpret or comment on the state of the science. Although he told reporters accurately that "All three of us"—Brix, Fauci, and I—"are very clearly against" opening up the country to reach herd immunity, Tony sent a very mixed message. He did believe the combination of vaccines and natural immunity could get us there, although he was unsure what percentages were required to get to "there." Early in the pandemic, he suggested that herd immunity could be achieved if 50 percent of Americans were vaccinated. Then he boosted that to 70 to 85 percent. Asked to explain how he had reached those estimates and why he changed his mind, he explained that he had based his guesstimate on the success of the measles vaccine, pointing out measles is the most easily transmitted virus. "When you get below 90 percent of the population vaccinated with measles, you start seeing a breakthrough against herd immunity," he replied, "people starting to get infected. . . . So I made the calculation that COVID-19 is not nearly as transmissible as measles."

He admitted he had changed his response to fit the evolving situation. "When polls said only about half of all Americans would take a vaccine, I was saying herd immunity would take 70 to 75 percent. . . . Then when newer surveys said 60 percent or more would take it, I thought, 'I can nudge this up a bit,' so I went to 80 to 85 percent . . . We really don't know what the real number is. I think the real range is somewhere between 70 and 90 percent. But I'm not going to say 90 percent."

The actual answer, I remained convinced and told anyone who asked, was that we would never achieve herd immunity. It did make me unhappy that Tony was selling what I believed was a false narrative. I understood his reasoning. He was trying to convince people to go get vaccinated as soon as it was available. But I didn't believe in selling false optimism. Herd immunity was the impossible dream. People who weeks earlier had never even heard the expression suddenly had become experts on it. It was going to lead us out of this pandemic. And too many people wasted too much time pursuing it.

Instead of pursuing this fantasy, the country slowly reopened, like a door creaking open to test the weather outside. Rapid testing, a result in fifteen or twenty minutes, finally had become widely available and allowed people to find out if they were infected. Offices, factories, and restaurants found ways of adding distance between workers or used plexiglass shields to protect customers and clients. Many cities allowed restaurants to add previously prohibited outdoor seating, sometimes in decorated sheds or booths, a strategy that kept numerous places in business. Food delivery options expanded. Weddings and celebratory events were held outdoors with reduced guest lists. Movie theaters, arenas, and stadiums allowed limited spectators to trickle in. Passengers wore masks in planes and trains. The big box stores insisted customers wear masks and improved ventilation systems.

A lot of it made sense, some of it had little protective value but made people feel safer, and other things were downright silly. For me, the biggest disappointment was how long it took schools to reopen. We had compiled strong data showing children were at little risk of life-threatening infections, and there were steps schools could take to mitigate the spread. Moreover, parents could use at-home tests to screen their children before sending them to school, and there was little evidence kids brought infections home to more vulnerable people. The CDC issued our first guidelines about safely reopening in May. We urged people, especially teachers and administrators, to use masks, to close cafeterias, to cancel extracurricular activities, and to implement ventilation, ventilation, and more ventilation. Congress had authorized $13 billion for schools to safely continue educating children.

I felt the damage that was being done far outweighed the consequences. In July, I told a Congressional committee that "Having the schools closed is a greater public health threat to the children than having the schools reopen. . . . It's important to realize that it's in the public health's best interest for K–12 students to get back into face-to-face learning. . . . 7.1 million kids get their mental health service at schools. They get their nutritional support from their schools. We're seeing an increase in drug use disorder as well as suicide in adolescent individuals. I do think that it's really important to realize it's not public health versus the economy about school reopening."

Overall, I think, I lost this battle. Too many schools stayed closed too long. By the time the country's public schools reopened, Joe Biden had been elected president, and I had been replaced.

In reality, whatever advice we offered was going to have only limited impact. The fact that the CDC is a government agency meant a significant portion of the population was immediately suspicious of our motives. The government was not going to tell these people what to do. People on the internet floated all types of absurdities, beginning with the absurd claim that the government had developed and was spreading Covid to give it an excuse to take total control of the country—martial law without declaring martial law. This conspiracy theory was only the first of a seemingly endless stream of ridiculous claims.

But there were a substantial number of people who believed them. Incredibly, there actually were some people in the administration who suspected the whole thing was a hoax. Deborah Brix suggested that doubt was triggered because "The information was confusing at the beginning. . . . They saw people get Covid and be fine, and then they had us talking about how severe the disease is and how it could cause these unbelievable fatalities." There were even more people infuriated by the entire situation; in some cases, they were mad at the government for not adequately protecting them. But what it came down to was there were a lot of desperately unhappy people who needed someone to blame for their lives being disrupted.

That was us. Politicians took the brunt of it, including Trump and later Biden. But as the people who had delivered what was mostly bad news, Tony Fauci, Deborah Brix, and I were punching bags. We were hammered almost every day for shutting down the country.

The fact—fact—that there was no one to blame for the creation and spread of a deadly virus did not stop people from blaming us. It was as if they believed we were responsible for it. Maybe it made them feel better. They couldn't blame Mother Nature or science, so they blamed the people trying to explain to them what was going on, people spending sixteen hours or more every single day trying to find ways to save lives.

I had weathered similar storms decades earlier while fighting the AIDS battles; but compared to this, those attacks seemed almost minor. For obvious reasons, the pandemic dominated every news cycle, even when there was little to report. So every word I said, every directive we issued, was scrutinized and too often criticized. I was criticized for shutting down industries. I was criticized for not opening up industries. I was criticized for not knowing the answers—even before we knew what questions to ask. Indeed, I was criticized for getting the science wrong. For example, before we had the data, I mistakenly believed Covid was similar to SARS and MERS. But that's how science works. You propose a hypothesis and, when the data emerges, you correct it.

I often got phone calls from very upset friends and family members complaining about me getting beaten up in the news. There was a real effort to discredit me; several stories dredged up the controversies from the AIDS epidemic, claiming completely inaccurately that I had proposed opening "concentration camps" to isolate patients. It was garbage, but it damaged my reputation. The media exploited any policy disagreements between members of the task force to cast doubt on our expertise and advice and to attract readership. As the competition for attention got more intense, the claims got wilder. After, when I said I believed Covid had been created in a laboratory in Wuhan, several newspapers accused me of being racist against Asians. Later, after I had retired and was working for free—free—for Maryland Governor Hogan, the state legislature voted to formally censure me and asked Governor Hogan to fire me. The governor refused and instead strongly supported me.

The difference, once again, was the internet. The internet made it easy for anyone to post fabricated stories, outright lies, and unsubstantiated opinions without any evidence to support them. A person may never meet the people they viciously attack, so there is little fear of consequences. We were attacked on both a professional and personal level. It sometimes seemed they wanted to proverbially kill the messenger for delivering the message. The president and members of his administration, including Azar and Atlas, tried to deflect responsibility for the situation by scapegoating the CDC, NIH, and FDA. I was directly criticized by the president several different times, and then I was criticized by people for not standing up to the president. Our motives were impugned by members of Congress. Too many politicians on both sides of the aisle engaged in simplistic forms of grandstanding, fanning the flames of their angry constituents. These politicians seemed more interested in political theater than actually doing something to solve their country's complicated problems.

Maybe the most disheartening, and perhaps most surprising, response came from other scientists who objected to me stating my opinion that the virus had originated in China. I hadn't expected that. I literally received death threats from prominent scientists after being interviewed on CNN. Death threats. In addition to phone calls, Joy and I received letters filled with white powder, which I suppose was supposed to make me think it was anthrax.

We all got threats. After Georgia's Marjorie Taylor Greene demanded Tony Fauci be prosecuted for supporting mask mandates, calling his actions "crimes against humanity," Fauci received additional threats on his life. Tony was under terrible duress and ended up having to go into protective custody. Is this how we treat our fellow Americans?

One evening, after CNN had done an especially dishonest report about me, accusing me of scientific misconduct and lying about the way I had handled AIDS, the president called me. "Well, Robert," he said, "if you can survive that, you can survive anything because they threw every bullet they had at you. We both know it isn't fair or accurate, but by attacking you they're trying to get to me. Just keep up doing what you're doing."

On another occasion, after my dispute with Atlas had gone public, he called to reassure me that he felt I was doing a fine job, and as far as my position was concerned, I had nothing to worry about.

Of course, those were personal calls. Private calls. A lot of those dangerous distractions could have been avoided if we had gotten more public support from the administration, but that didn't happen. I think it is fair to say that in terms of using science to create political policy, science is an inexact science. Science is a staircase of progressive testing: What is true in the lab often fails in human testing. Furthermore, data can be misleading, misconstrued, and manipulated. Mark Twain once famously said, "there are lies, damned lies, and statistics." As I saw during the AIDS crisis, data can be presented in ways to support the point you already want to make.

In the middle of a reelection campaign, the administration needed good news. It had to convince voters it was in control of the pandemic. That it wasn't as bad as people like Redfield were saying, that it would be over sooner than the gloom-and-doom people were saying. President Trump at times would make upbeat public statements that couldn't be factually supported. There were times he directly contradicted me, adding to the confusion. But he also deserves great credit for taking the necessary steps to meet this challenge. It was President Trump who initiated and funded Operation Warp Speed, which resulted in a vaccine being produced more rapidly than ever before in history. It was the president's leadership that resulted in the Covid vaccine being available for widespread distribution to the American public the day it received FDA approval. That decision saved thousands of lives. It was the president who funded billion-dollar grants to mitigate the spread. And where I was concerned, he was mostly supportive.

Meanwhile, day after day after day, we watched the death toll growing from tens of thousands, to hundreds of thousands, to eventually more than a million Americans, while simultaneously tracking the race to produce vaccines.

During that first year, I had gotten to know the enemy quite well. That's also science: When you see or are confronted by something new or interesting, you begin by making assumptions based on what you already know, then you start testing them. Our initial belief that SARS-CoV-2 was another form of SARS or MERS was proven wrong. Our assumption that it was spread when people were symptomatic was also wrong. Since then, countless studies on the path to a vaccine have revealed the characteristics of this novel coronavirus.

Covid actually is considerably different than either SARS or MERS. About 80 percent of its genome sequence is similar to SARS and only 50 percent is the same as MERS. In fact, the closest relative to Covid is a virus found in bats in the Yunnan Province of China, which is a bit more than 90 percent the same. That is one reason so many people suspect Covid came from bats.

We learned that people of all ages are susceptible to infection, but the median age is about fifty years old. And, as is predictable, older people with comorbidities are the most vulnerable. We learned how it attacks the body, what the symptoms are, and that those symptoms can persist for two weeks or even longer. We learned that in addition to the expected gastrointestinal effects, it also can cause kidney and heart problems. We learned that asymptomatic transmission accounted for a majority of new infections, that this virus can hang suspended in the air for a considerable period of time, that, in addition to human-to-human transmission, it can live on surfaces—and while it can survive for as long as a week on hard surfaces like glass, plastic, and metal, it was shown to remain alive on paper for a half hour and on some fabrics for two days. We also learned that something as simple as soap and water kills it.

We released updated guidelines every day. Piece by piece, protein by protein, we figured it out.

Until we didn't.

To our surprise, we discovered that for some people the symptoms didn't just go away. They persisted for weeks, in some cases for months, and, in rare cases, for years. We hadn't seen much of that in SARS or MERS. That doesn't mean it didn't exist, just that science was not aware of it. This syndrome actually was identified by patients, most of whom found fellow sufferers, rather than physicians. In May 2020, it was described as "long Covid" for the first time. It came to be defined as the persistence of symptoms far beyond the normal time range of the disease.

It's an unusual, but quite common, condition. The number of people who develop this after an initial infection ranges from 7.5 percent to as many as 20 percent. Its symptoms vary widely but can include brain fog; muscle weakness and dysfunction; mental impairment; shortness of breath; joint pain; temporary loss of smell and taste; and even the loss of speech. But because these symptoms did not show up in standard medical testing—the usual diagnostic tools like metabolic panels and blood work—physicians generally attributed them to a range of mostly psychological problems, including depression, anxiety, "women's trouble," a mental breakdown, and drug use. As one patient explained in a study, "In the beginning, it was terrifying. No one believed or understood that covid lasted longer than 2 weeks. . . . At the most terrifying point of my life I had to fight not just to live but for people to believe that my illness existed let alone get help . . . because 'covid is only respiratory and lasts max 2 weeks.' There are people like me who survived and live (exist because this is not living) in a haze but have never been back to themselves."

Doctors who denied this condition existed, because they could not find physical evidence, were accused of "gaslighting" patients. Long Covid was actually illuminated on social media before it became a recognized medical condition. Patients found each on the internet, compared symptoms, traded treatments, and slowly grew and organized. In late 2020, the World Health Organization finally acknowledged it was real.

Since then, there has been considerable research, but it still remains a mysterious condition. We don't know how to prevent it. The severity of the original infection doesn't seem to matter. People who had a mild case can suffer as much as those who had a severe case. There still is no treatment. In many cases, it simply goes away, although it may reoccur. In severe cases, it can last for years.

Several months after leaving the CDC in January 2021, when Joe Biden became president, I became a consultant to Maryland Governor Larry Hogan and also resumed my clinical practice. I loved that; seeing patients is the front lines of medicine. For me, it also opens the door into practical research. I began seeing patients suffering from long Covid. A lot of physicians were not interested in taking care of them; they continued to believe it was a psychosomatic condition. There were no recommended treatments, and nobody knew what to do for these people. So in numerous cases, these physicians referred their patients to psychiatrists.

The situation was very similar to those early days of my career at Walter Reed, when young men and women were coming to the hospital with AIDS, and nobody wanted to treat them. I suddenly was confronted with a significant number of people whose medical issues were not being met. Even then, I knew enough to reassure them: You're not crazy, you don't need a psychiatrist. They had an illness, post-Covid syndrome was how I described it. It was long Covid.

There no longer is any doubt long Covid exists. I learned firsthand how brutal it can be. What struck me was the prevalence of cognitive dysfunction. I began to think of it as an acquired Alzheimer's. The only difference is, as opposed to Alzheimer's, it gets better. The number of people suffering from this turned out to be far greater than anyone realized. I estimate as many as 15 million Americans have suffered from this impairment. I could be wrong; it may be far more than that.

I treated some extreme cases. Among them was a forty-seven-year-old professor at one of Baltimore's major universities; she literally lived on a stretcher for almost six months because she couldn't control her pulse or blood pressure if she sat up. I had patients who lost their ability to speak in complete sentences longer than two or three words; they couldn't remember words, including their own names. I had patients with severe breathing problems because the valves in their veins stopped working properly, disrupting blood flow to their heart. The variety of symptoms was astonishing. There was no predictability and, unfortunately, I had no real treatments to offer.

I've done considerable research, trying to figure out what causes this and what might help my patients. It is not clear if the culprit is the virus or proteins left behind.

All my original AIDS patients died, while all my long Covid patients survived. Time is the cure. But because the symptoms are so varied, there is no standard treatment. My objective became minimizing the symptoms while reducing the time they persist.

I had done a significant amount of experimental clinical research with both AIDS and hepatitis. The common and dominant complication of long Covid is microcoagulation; Covid actually is a blood vessel disease that results in the patient's blood clotting. So I started treating my patients with three different anticoagulants, a type of regime initially suggested by a South African physician. A doctor in Birmingham, Alabama also was experimenting with it, and we regularly compared results.

Physicians are permitted to use drugs that are legally approved for other uses when they believe there is a good medical rationale for it. The best-known example of that is sildenafil, the hypertension drug that became Viagra. The results in some cases were remarkable. Combined with reassurance that long Covid is a physical condition, not a psychological problem, and that it eventually goes away, the course of treatment with anticoagulants overall has been positive. That said, people continue to get long Covid. It persists.

As does COVID-19. It hasn't gone away. The pandemic has ended; but, like the seasonal flu, the virus remains a threat. People don't talk about it much, but by 2025, it continues to kill about two hundred Americans every week. And it is here to stay. Meanwhile, we have started analyzing mountains of data, trying to figure out what we learned from the steps we took in dealing with the pandemic. The response to the pandemic caused three significant societal disruptions: the debate about mandating masks, the creation of remote offices, and the closing of schools.

What worked? What saved lives? What failed? There is absolutely no question that taken together these policies saved tens of thousands of lives. While it is impossible to determine any specific figures, the *Journal of the American Medical Association* reported that Mississippi, which imposed the least restrictions, had five times the per capita death rate of Massachusetts, which maintained strict adherence to the guidelines. The question remains: What value did each of the three primary interventions have?

While they remain controversial issues, with support or disregard generally divided by political loyalties, there just shouldn't be any debate that wearing masks saved lives, tens of thousands of lives. Beyond the simple logic, spewing fewer particles into the atmosphere results in fewer infections, the vast majority of numerous studies have supported it. While these studies take different approaches and measure different outcomes, and while they provide different estimates of how many lives might have been saved, they reach the same conclusion: Mask mandates saved lives. For example, the NIH published a study in February 2023 reporting that between January and December of 2020 "statewide mask mandates saved 87,000 lives through December 19, 2020, while an additional 57,000 lives could have been saved in the same period if a nationwide mandate had been enacted starting in April 2020."

In a focused study of Kansas, a state in which large areas resisted mask mandates, the numbers were telling. In the weeks following the imposition of a mandate, infections declined significantly, and the reduced number of new infections reflected the adherence to that mandate. A substantially higher percentage of people were hospitalized or died in areas that refused to wear masks than in those areas where people wore them.

An extreme comparison showed that by August 2020, New York City, which had a mixed response to wearing masks, had suffered almost 40,000 deaths. Meanwhile, Hong Kong, a city roughly the same size as NYC that had an almost universal adherence to mask restrictions, reported only sixteen deaths.

A 2023 systematic review by Britain's Royal Society reported that masks reduced risk by as much as 18 percent, while the *British Medical Journal* claimed masks reduced chances of infection by 25 to 71 percent. A meta-analysis published by *America's Journal of the American Medical Association*, which analyzed the value of mandating masks, concluded that studies in "more than 400 US counties showed that enactment of a mask mandate was associated with a 25 percent reduction in COVID-19 incidence 4 weeks later."

The value of remote work in terms of saving lives during the pandemic also should be debatable. Less social interaction obviously results in fewer infections. That said, the effect on the economy, production, and personal relationships is a lot more difficult to measure. While other protective measures mostly disappeared with the end of the pandemic, working from home has become an accepted practice. It worked. It worked so well it has become ingrained in the corporate world.

The question becomes: Are the individual benefits as important as the professional impact? There are studies to support almost any possible conclusion. Eliminating the commute to an office, institutionalizing flexibility, and allowing people to spend considerably more time at home is enormously popular with workers, while being able to downsize office space and materials has resulted in significant savings in operating costs for companies, and it has allowed employers to greatly increase the potential talent pool and to reduce absenteeism. But all of this has come with a cost. The loss of face-to-face relationships, reductions in corporate morale, the loss of camaraderie and often-productive water-cooler conversations, the lack of mentorship and supervisory opportunities, and inherent psychological problems like isolation and a blurring of work-life structure have become unfortunate staples of our new economy.

It's impossible to make any definitive conclusions. Remote work worked. It saved lives, and it isn't going away. While studies seem to suggest that, rather than decreasing productivity, it has increased it, some larger companies have reverted to a pre-Covid traditional five days a week in the office schedule. Others are experimenting with a hybrid workweek.

Like remote working, closing schools also has had a lasting effect, but the long-term outcome seems to be a lot less positive.

All fifty states closed their public schools in March 2020. I always argued strongly against it. This wasn't the 1918 pandemic. During that period, in which young people were at risk, the closure of schools for weeks, and in some cases for months, did make some difference in infections and deaths. It saved a limited number of lives. And it did not appear to have much of a negative impact on their future.

By contrast, we learned early that Covid was nowhere near as dangerous to young people as the Spanish flu. We also had developed significantly better methods of protecting them in classrooms. I was a lot more concerned about the academic and psychological damage than physical ailments. In some areas of the country, public schools were closed for more than a full school year. Subsequent testing has demonstrated that remote learning negatively affected academic progress. A 2023 analysis of test scores from 7,800 school districts compiled by respected educators concluded that "pandemic-related learning losses are historic in magnitude." Compared to the previous three years, "In districts where students spent most of the 2020–21 school year learning remotely, they fell more than half a grade behind in math on average, while in districts that spent most of the year in person they lost just over a third of a grade."

An ongoing debate concerned how long should schools remain closed. That decision was left entirely to state governments. But in July, the CDC issued guidelines strongly urging districts to open as quickly as their schools could be made safe. As I said, "It is critically important for our public health to open schools this fall. School closures have disrupted normal ways of life for children and parents, and they have had negative health consequences on our youth. CDC is prepared to work with K–12 schools to safely reopen while protecting the most vulnerable." And yes, I responded when asked by reporters, I absolutely would send my grandchildren to school in the fall.

There is no question keeping schools closed for a prolonged period hurt students. A Harvard analysis concluded, "In districts that went remote, achievement growth was lower for all subgroups, but especially for students attending high-poverty schools. In areas that remained in person, there were still modest losses in achievement, but there was no widening of gaps between high and low-poverty schools in math (and less widening in reading)."

Data collected in the ensuing years has continued to show that students have not recovered from the academic damage done by closing schools—especially students in the lower grades who should have been learning fundamental reading, writing, and math skills in person. Students also lost their version of water-cooler socialization, leading to widespread behavioral and mental health issues, problems that have continued to surface even as the students mature.

Numerous factors have made it extremely difficult to compare educational outcomes between states that opened early and states that remained closed; that includes everything from the local weather that permitted classes to be conducted in fresh air to the method used to count Covid deaths. But the general consensus is those states who kept their schools open as much as possible did seem to have a marginally better educational outcome. That said, as much as it is possible to compare death rates of states that reopened quickly to those that stayed closed, there doesn't seem to be identifiable patterns. A 2025 study of five countries in various parts of the world concluded, "reopening schools did not change the existing trajectory of COVID-19 rates."

During that first year, we employed every weapon we had against this virus. People stayed home. In banks, and shops, and restaurants, they maintained a six-foot social distance. Masks saved lives. Remote working had benefits that continue to reverberate. We closed schools. There is little evidence that closing schools saved lives, but it did cause lasting harm to students that will affect them in the future.

We were doing everything that seemed to make sense, anything that prevented people from making physical contact, until we had the weapons we needed to defend ourselves.

Until we had a vaccine.

NINE

A vaccine is a promise: It will shield you, it will protect you, it will ward off the enemy. And like most promises, there is often a degree of exaggeration in it. A vaccine is never a guarantee. It won't protect every person who is inoculated, and most vaccines are not durable. But even knowing that, I believe completely that vaccines are the miracle that science has given to humanity.

It is beyond impossible to estimate how many tens of millions of lives vaccines have saved since Jenner's 1796 smallpox vaccines. I've spent my entire life getting poked with vaccines. I was vaccinated as a baby against the range of childhood diseases. When the new polio vaccine became available in 1955, I stood in line with my classmates, closed my eyes, and got that protection. When I was inducted into the United States Army, I think I got seventeen vaccines in one day. I continue to get my annual flu vaccine. When the Covid vaccine became available for my age-group, I didn't hesitate to get inoculated, and I subsequently have gotten several additional shots. Joy, our children, and our grandchildren have all been vaccinated. I have seen vaccines working under a microscope, I have spent days and weeks analyzing data. I *know* vaccines work.

Most of the time.

I know that the power of vaccines to protect people is sometimes overestimated, and I know that serious mistakes were made with the introduction and distribution of the COVID-19 vaccines.

There was no flu vaccine during or even after the 1918 pandemic. Before that time, several vaccines—weakened or inactive doses of a pathogen that trigger the natural immune system to produce antibodies to kill the invaders—had proved effective in fighting rabies, typhoid fever, and diphtheria. But scientists mistakenly believed the Spanish Flu was caused by bacteria, bacillus influenzae. Worldwide efforts to create a vaccine failed because they were focusing on the wrong pathogen.

In 1933, researchers in England, who were examining fluids extracted from the nose and throat of flu patients, did not find any bacteria; instead, they were able to isolate and identify the virus that caused this disease. The researchers, who had used ferrets to prove their theory, reported cautiously, "The evidence strongly suggests that there is a virus element in epidemic influenza, and we believe that the virus is of great importance in the etiology of the human disease."

Dr. Thomas Francis, working at New York's Rockefeller Institute confirmed that discovery. He also found that viruses were capable of changing their chemical structure to survive, that they were devious pathogens that literally changed their composition to evade specific human immune cells hunting for them. That meant it was impossible to protect against influenza with a single vaccine. Instead, updated vaccines would have to be produced to target the specific strain of the evolved virus.

Francis began pursuing that vaccine at the University of Michigan. Among the young scientists he recruited to work with him was a twenty-seven-year-old virologist named Jonas Salk. Salk was put in charge of the day-to-day pursuit of the flu virus. By the Fall of 1942, Francis and Salk were ready to test their rudimentary flu vaccine. At that time, lax regulations allowed researchers to secretly test new drugs on uninformed, unsuspecting patients. The military, having seen its forces decimated by influenza during the Great War, facilitated this test.

At the beginning of the 1942 flu season, blood samples were taken from eight thousand mental patients, who were then given the new vaccine. A second blood sample taken two weeks later showed flu-fighting antibodies had increased about 85 percent.

Months later, a double-blind trial was held. This time, one hundred male psychiatric patients were vaccinated while an additional one hundred men received a saline placebo. Then all of them were exposed to the flu virus; a nasal mist carrying the virus was sprayed into their nostrils. They literally were given the flu. That's the way medical research and testing was done. Once again, the results were tantalizing.

Prior to the next flu season, 12,500 army trainees unknowingly participated in a massive double-blind test, which this time included the two known primary serotypes, or strains of the disease. The results changed the world. Only 2 percent of those troops who had been vaccinated had gotten the flu. But equally important, the data told Francis and Salk that within those communities significantly fewer unvaccinated people came down with the flu. The protection had somehow been extended to them. This was the first time the power of this so-called "herd immunity" had ever been proven.

The long-sought protection from this potentially deadly disease had been found. The entire army was vaccinated, and only 8 percent of those soldiers eventually got sick. A few years later, Salk adapted similar laboratory techniques to develop the polio vaccine. This time though, it was tested first on monkeys and then Salk, his family, and several volunteers. When it proved to be completely safe, the largest double-blind clinical test in American history took place: More than 1,000,000 students were inoculated.

The vaccine proved to be safe and effective. Several decades later, polio had essentially been eradicated.

Flu vaccines work, but the tricky part is developing a vaccine against each newly evolved influenza strain. Each year, months before flu season began, researchers have to make an educated guess about which strain will be prevalent before they produce a vaccine. The process takes at least six months, and even then, the effect of the vaccine is limited. For a variety of reasons—including how closely the vaccine matches the evolved strain, the number of people who choose to get vaccinated, and their underlying conditions—rarely does the vaccine provide protection for more than half the population. Some seasons, when researchers guess wrong, it protects 25 percent or less. The year I became director of the CDC, the strain we used protected only 23 percent of people who were vaccinated.

Unlike the influenza vaccine, the polio vaccine provides essentially permanent protection because there are only three different serotypes; the Salk and Sabin vaccines contain a mixture of all three. So, the virus has no place to gain a foothold in your body. The measles vaccine offers similar immune response.

That's the reality we were dealing with when we launched Operation Warp Speed. At best, the vaccine would offer only limited protection. Few health professionals anticipated one vaccine would be able to eradicate the SARS-CoV-2 virus; but combined with other mitigating factors, it would enable us to end the pandemic.

Operation Warp Speed certainly is one of the most successful vaccine development programs in history. To be able to put shots in arms in less than a year is astonishing, and it marks the beginning of a new age of preventive medicine. But as we later learned, that was probably not the best name for the project. It suggested science fiction rather than good modern science. The speed at which this was accomplished raised questions about how it possibly could have done so quickly without compromising safety. And it provided rhetorical ammunition for vax skeptics.

We started with a platform that had been developed to combat SARS and MERS, and then we plugged the genome released by China into it. We had our vaccine candidates. Now we had to do everything else.

We maintained scientific integrity at every step. Traditionally, the development of a new vaccine begins with a phase 1 study in which a small number of people are inoculated. Researchers collect data on how safe it is, its side effects, and its efficacy. Does it stimulate an immune response? That data is analyzed. If the vaccine works, and it is safe, it goes to stage 2. This is a greatly expanded trial in which hundreds of people, often a variety of ages and ethnic backgrounds, get the vaccine. That resulting data is analyzed for safety, potential risks, and the ability to produce an immune response. Phase 3 includes thousands of participants who are monitored for any anomalies, side effects, and unusual reactions—any potential stop signs. We evaluate the success of the vaccine in respect to what it is supposed to be doing. In certain situations, the company might conduct a second Phase 3 study so they can confirm the vaccine's efficacy. During this stage, the FDA also examines the proposed manufacturing facility. If the vaccine proves safe and efficacious, the FDA can approve it for Phase 4, in which it is in use by the general public while additional data is collected, often over several years.

That process can take five years or longer. Previously, the quickest a vaccine had gone from development to implementation was the mumps vaccine in the 1960s, which took four years.

We didn't have years. We were in the middle of a pandemic. Thousands of people were dying every day. We were in the save-as-many-lives-as-possible phase. To save months, we blended Phases 1 and 2 to get initial safety and immunogenicity data. No time was wasted. Steps overlapped. The FDA made this its priority. And every piece of data was immediately analyzed. Once we knew it was safe and it worked, we allowed six companies to start manufacturing the vaccine, funding that process by committing to buy 100 million doses after the FDA approved it for human use. Moving forward with six companies allowed us to kind of hedge our bet. We weren't going to wait to finish the study to begin the next round. Obviously, it was risky. If the vaccine didn't work, we were going to have a lot of little vials as souvenirs. It was a gamble. We maintained strict safety protocols at every step. We won that gamble.

But having a viable vaccine and successfully vaccinating the American public were different challenges. This pandemic was not going to spontaneously end. It was going to end because of the biological countermeasure known as a vaccine. But that vaccine needed the American public to take it. We weren't going to shut down this pandemic if 30 percent of the American public refused to take it.

Unfortunately, the Covid vaccine—even before it was approved—became mired in political debates. Questions were raised about its actual value in preventing the disease, the potential dangers, and whether people should be forced to take it.

Science went to war with opinion. Once again, the country was split roughly by political beliefs.

Even before the vaccine was available, we were debating how it would be distributed. First responders and essential workers certainly were a priority. Hospital workers had to be vaccinated. The teachers' unions wanted their members to be vaccinated so the schools could safely be reopened. The elderly in nursing homes were the most at risk. There was no right answer. Different states had different priorities. New York, for example, had few functioning firefighting units because so many of their people had been isolated. Finally, the administration decided to leave that decision to the governors.

But before distribution even began, skeptics raised doubts about the efficacy of the vaccine. Rather than being celebrated, the timing made people suspicious. There was no question that having a vaccine before the presidential election in November 2020 could help the Trump campaign, and that made some people doubt it was real. The politicization of the vaccine began in early September when Democratic vice presidential candidate Kamala Harris told reporters, "I would not trust Donald Trump" when it comes to the reliability of the vaccine "because he's looking at an election coming up in less than 60 days, and he's grasping for whatever he can get to pretend he has been a leader on this issue when he has not."

Claims that the pharmaceutical companies were playing politics with the vaccine seemed to be reinforced when Pfizer announced that in clinical trials its vaccine had demonstrated 90 percent efficacy—one week after the presidential election. Conservative activist Charlie Kirk spoke for a lot of people when he questioned the timing of that news "Isn't it interesting that probably the best news for the market and for potential swing voters that voted a week ago was just released this morning?" he asked. "Regardless of your views on vaccinations," he continued, "I think it is widely agreed upon that the news of an effective vaccine would help the markets and also help President Trump." Although Pfizer's CEO completely denied this, for a lot of people the vaccine had become a political tool.

The fact that it had been created faster than any vaccine in history and was based on a new technology raised additional questions. Had sufficient clinical trials been conducted? Is there any way to know what effect it might have five years in the future? Ten years? A growing online movement skeptically began claiming the vaccine was "experimental," "untested," and even dangerous.

The vaccination program began on December 14, when Sandra Lindsay, the director of critical care nursing at Long Island Jewish Medical Center, was inoculated. "It feels surreal," she said, "it is a huge sense of relief for me, and hope." It was televised live on CNN.

All of the members of Warp Speed and the task force took tremendous pride in that accomplishment. We had done something no one had ever done before. And while it was just the very beginning, for me personally it was the beginning of the end of my tenure at the CDC. President Biden had won the election. I would have been happy to serve in his administration and, being in the middle of a pandemic, I thought it made sense. Instead, they decided to replace the entire team. That's the way the system worked. We opened up the CDC to the new people to get them up to speed.

I probably didn't realize how large a toll my term had taken until it ended. It took me a few months to decompress. I learned that I had diabetes II, and my AIC, the measure of the disease, had shot up from a very manageable 5.8 to a terrible 24. While the reasons for that are complex, stress is a primary factor.

And the stress had been unrelenting.

When I accepted the CDC position, I chose not to take a leave of absence from my posts at the University of Maryland, among them vice chairman of the Department of Medicine and head of infectious disease research. I resigned to eliminate any question of conflicts of interest concerning grants going from HHS or CDC to the university. But as a result, I didn't have a job.

I started a private consultancy business and became an unsalaried senior advisor for public health to the state of Maryland and Governor Larry Hogan. It was an intriguing transition; I was now on the other side of the pandemic. I had gone from developing the vaccine to putting it into use in the field.

At my suggestion, the state began by vaccinating nursing home patients. By the end of February, we had them nearly all vaccinated, and we immediately saw a huge drop in the number of individuals who needed to be hospitalized. I was watching everything I had done at the CDC put into action. But in late April and May, the data told me a new story: The number of nursing home residents getting sick and having to be hospitalized was rising. I told Governor Hogan, "I don't think this vaccine is lasting. I don't think it's durable." We tested about 750 people who had been vaccinated; almost two-thirds of them no longer had evidence of an immune response. Meaning the Covid vaccine acted more like the annual flu virus than the enduring polio virus. That was not really a surprise. "We need to revaccinate them," I insisted.

The CDC, NIH, and FDA all told us that was not necessary. That was a real problem because the federal government had paid for these doses and was distributing them to the states. There still was somewhat of a shortage. Maryland was told it could not revaccinate its nursing home patients. But I pushed, and Governor Hogan agreed with me, we had to give those residents a second shot. Literally, in many cases, a second shot at life. We did and the numbers again dropped. We had compiled good empirical data that the vaccines last only between three and six months. It took several months before the CDC officially recommended booster injections for vulnerable individuals.

That response infuriated me. The government—and until a few months earlier I had been part of it—had taken the very authoritative position that it knew better than we did. It decided that people were going to be vaccinated on a schedule set by the government. Well, I knew how to interpret the data, and I now knew what people were being told wasn't necessarily accurate. I am not—not—questioning anyone's motives. There are many factors for this recommendation. But it was the attitude—we know better, don't question us—that upset me. It was the beginning of more controversies that plagued the entire vaccination effort.

Mistakes were made. Again. But probably the single most significant mistake was letting people believe that once they got the vaccine they were protected against Covid. That wasn't true. That was never true, But that was the impression sold to the American public. We learned very quickly that the vaccine did what it was approved to do: It prevented people who were adequately vaccinated from getting seriously ill, getting hospitalized, and dying. But the vaccine didn't prevent a milder infection. It also wasn't a permanent solution. Many people who had been vaccinated were stunned when they caught Covid a second time. The fact that it did not prevent transmission went a long way toward eroding confidence in anything the government said about the vaccine.

It shouldn't have been a surprise to those who knew the science. It had long been evident that the natural immunity—the antibody response developed after exposure to the virus—supplied protection for only a limited period of time. Some people had already suffered through two or even more cases of Covid. So there was little reason to believe the response generated by the vaccine would be significantly different.

Then the American people were told they needed two shots to be fully vaccinated. That was the phrase that was used, fully vaccinated. Two shots. Then I'm done. I'm fully vaccinated.

Except that also wasn't accurate. The Covid vaccine didn't work like that. The vaccine does not prevent infection. It prevents serious illness and death. I was so-called "fully vaccinated" after I left the CDC. I caught Covid twice. Maybe I didn't feel great, but I'm confident those vaccinations kept me out of the hospital. The truth is we still don't know enough about the durability of this mRNA vaccine. Unlike traditional vaccines, this one turns your body into an mRNA factory, and we have no control over production. We don't know how to regulate it. That factory can churn out the vaccine for a week, or it can make it for six months. But now, in addition to the mRNA vaccines, we have a traditional protein vaccine made by Novavax. That vaccine uses dead proteins to stimulate your immune system. Your body gets what it gets, the factory never opens. Although I know the mRNA vaccines fulfill their purpose, the last several times Joy and I have been vaccinated we've gotten that Novavax protein vaccine, and that is what I now give my patients.

The flames of doubt fanned by that misunderstanding of the science made it easy for the growing anti-vaccine movement to take hold and grow. There already was a receptive audience for it. The existing anti-mask groups found a whole new cause—the efficacy of the vaccine—and adopted it. The internet allowed people to find each other and organize into large and somewhat powerful groups.

In addition to the most common misconception—the vaccine had not been sufficiently tested and rushed into production without proof it was safe—all kinds of other conspiracy theories started spreading on the internet. Supposedly Bill Gates, Fauci, and other people had created Covid to force people to buy their expensive vaccines, that manufacturers were more concerned about selling vaccines than keeping people healthy, that the vaccines actually contained an electronic chip that enabled government to track people, and that this was only the first step towards taking away our freedom and liberty.

The existence of anti-vax movements is as old as vaccines. They probably began a day after Jenner announced his smallpox vaccine. Some people have always been very uncomfortable being given a dose of a dangerous disease, even when they are assured it is safe. More recently, the publication by respected medical journals of studies claiming vaccines caused serious medical conditions like autism scared a lot of people—even when those studies were subsequently discredited.

The situation was complicated by Americans' growing distrust of the pharmaceutical industry. While generally people acknowledge these companies do a good job developing new drugs, only about half of the population believes their claims about their products. There is a thin line between medicine and marketing.

The result was that millions of people didn't know what to do and vaccine hesitancy took hold. It made no difference that this vaccine did not contain a sample of the virus. Instead of getting themselves and their children vaccinated, it was estimated that more than 20 percent of Americans chose to wait or made the decision not to get it. In some states, among them Wyoming, Louisiana, and Mississippi, almost half of all residents refused to be vaccinated. Rather than directly addressing the problem, individual states, private industry, businesses, and eventually the federal government adopted the worst possible solution: They mandated vaccination. They forced people to get the shot.

While that arguably prevented some people from a potentially serious illness, what it also did was magnify the existing fear that this was some sort of government plan to take away personal freedom and further divide the country. It fed right into the spreading fear that government and industry were taking away freedom and liberty.

Vaccine hesitancy was already a serious and growing problem when I became CDC director. My first year in that position we had more children die of influenza than had ever before recorded. And when I looked at the data, I saw that the one thing they all had in common was they weren't vaccinated. Many of those lives might have been saved if those children had been vaccinated. But something was preventing their parents from getting it done.

The rate of infection and serious illness was especially high in minority communities. So even before Covid, I knew this was a big problem. One of my major objectives became fighting that vaccine hesitancy. I started a program designed to teach Americans how to be vaccinated with confidence. We worked hard on it, focusing on the Black and Hispanic media. We had some success, and we were able to raise vaccination rates in those communities. We didn't solve the problem, but we were making inroads when Covid hit.

What emerged from that program was an understanding of why people weren't getting vaccinated, especially children. I didn't realize at that time how important that knowledge was to become. The reason they resisted vaccination, I learned from talking to parents, was that they were scared of making a mistake, scared that something might happen to their children. They'd heard stories. There were rumors. They just didn't have enough information.

The answer, I found, was to provide the accurate information they needed to make the decision. Usually, if I sat and talked with them and explained the advantages of the vaccine, they were reassured. Take the human papillomavirus vaccine, for example, which is recommended for children at about ten or eleven years old. I asked parents, "Would you like to vaccinate your child against cancer?" The answer was always, oh yes. I explained, "The human papillomavirus causes cervical cancer for women and oral cancer for both men and women. This is a vaccine that can prevent your child from getting those cancers."

The response was overwhelming, "I want my child vaccinated now."

The same thing was true when I spoke with parents about the measles and rubella vaccines. The tremendous success of the measles vaccine means that very few parents know anything about it. They didn't get it, and they haven't seen it. When I explained to them that measles can cause permanent neurological damage and that in other parts of the world it is deadly, they generally were stunned, responding, "Measles can kill? I thought it was just a rash and a fever." That conversation most often led to them embracing the vaccine.

Another group I encountered actually was misinformed. When I asked them why they were not vaccinated, they would tell me something that was totally inaccurate, such as that the vaccine would change their genetic information. It wasn't fear that was stopping them, it was their belief in something that wasn't true. In those situations, I would try to correct them; sometimes I would give them material to read. And in many cases, they eventually changed their minds.

For some people, the problem was logistics. They couldn't afford to take time away from work to get vaccinated. Or they lacked transportation.

There was a small group that were just 100 percent anti-vaccine, and nothing I could say would make any difference. They were convinced vaccines were a substance that they should not put in their body. They believed that the potential cure for a disease actually was worse than the disease.

Part of my program at the CDC involved urging physicians to spend additional time with patients who refused to be vaccinated or have their children vaccinated. Rather than just accepting it, I wanted them to find out why those people felt that way. Talk to them honestly. If they don't agree, talk to them again the next time they came in. People want to be healthy, and they want their children to be healthy. The vast majority of people, I told them, ended up accepting vaccines once they understood their benefits and safety.

I couldn't emphasize it enough: Vaccines save lives. Vaccines prevent people from getting seriously ill, which can have devastating economic consequences. I used every argument I could to spread the word to physicians and later my patients.

So I was ready for the anti-Covid vax movement.

Or so I thought.

What really disappointed me was the government's response to vaccine hesitancy and outright rejection. If the government had been completely candid with Americans, the rollout would have been far more successful. It remains a tremendous mistake to claim that all people who refused to be vaccinated were making some kind of political statement. What many, or even most, of them wanted was for themselves and their families to be safe. But they were getting conflicting information. They were confused about what they were being told. It began with the reports that this was a virus unlike anything we had ever seen. Unlike other viral diseases, you could have Covid, and you could even infect other people, but you could have no symptoms. That difference itself was scary.

Then, months after being told they were safe after getting a shot, people found themselves standing in lines for a second shot. And sometimes a third shot. The fact that three different pharmaceutical companies produced different vaccines, which were rumored to have different properties, further confused people. Leaving it to states to decide which groups would get it in what order led to mixed messages. Next, the government ignored the fact that people who caught the disease had limited immunity and did not need to be immediately inoculated—although eventually it was suggested they could wait ninety days. And then people started learning about the different variants: the South African variant, the Brazilian variant, and the UK variant. In response to evidence that the vaccines had decreased efficacy against some of these variants, the companies began producing a booster, and people were told they needed yet another shot.

When President Biden was elected, he made it clear he was against a vaccine mandate, saying, "I will do everything in my power as president to encourage people to do the right thing and when they do it, demonstrate that it matters." But eventually that changed. When a new wave of the virus threatened the country in the fall of 2021, his administration issued sweeping mandates, forcing most federal employees to be vaccinated. In response, a federal judge issued a preliminary injunction that briefly prevented most federal employees from being vaccinated. The health issue had become a legal fight.

America has a long history of vaccine mandates, beginning in 1777 with General George Washington's Continental Army. General Benedict Arnold's march on British forces in Quebec, the first major offensive of the Revolutionary War, had ended in retreat when too many of his troops were lost to smallpox. In response, Washington told his army's chief physician, "Finding the smallpox to be spreading much and fearing that no precaution can prevent it from running through the whole of our Army, I have determined that the troops shall be inoculated. This expedient may be attended with some inconveniences and some disadvantages, but yet I trust in its consequences will have the most happy effects. Necessity not only authorizes but seems to require the measure, for should the disorder infect the Army in the natural way and rage with its usual virulence we should have more to dread from it than from the Sword of the Enemy."

This mandate was very effective; smallpox infections were reduced by more than two-thirds. Through the following centuries, as vaccines for numerous diseases became available, they were mandated for military personnel. Civilian mandates were very different. It appears individual states began compelling residents to be inoculated against smallpox in the late 1800s. People who continued to refuse could be fined. In isolated cases, they could even be imprisoned. Naturally, some people objected and initiated legal action. As one anti-vaxxer told the court in 1900, he refused to "have his system so poisoned by the vaccine virus as to result in his permanent injury."

Various courts wrestled with this question. But in 1905, the Supreme Court decided that a Massachusetts law that fined people who refused to be vaccinated was constitutional, that "it is within the police power of a State to enact a compulsory vaccination law."

The COVID-19 mandates unleashed a torrent of lawsuits. The first two cases to reach the Supreme Court were decided in January 2022. It was a split decision, in some ways mirroring the existing confusion. In the first case, the court found the government could require employees working at medical facilities serving patients enrolled in Medicare and Medicaid to be vaccinated. But in the second challenge, the court found a government agency, the Occupational Safety and Health Administration, whose mission was to ensure safety in work environments, could not force employees of most large corporations to be vaccinated. While not a single justice questioned the right of the government to issue vaccine mandates in specific situations, the court also decided Congress had not granted those sweeping powers to OSHA. The general rule seemed to be that government agencies, like HHS, could issue mandates only with congressional approval.

That has not stopped the lawsuits though. It's impossible to know the number, but a reasonable guess is that hundreds of lawsuits have been filed about the mandates, and many of them are still winding through the system.

Many other countries also instituted mandates. In January 2021, Indonesia became the first country in the world to essentially order residents to be vaccinated and penalize them if they refused. When the Austrian government announced on February 1, 2021, that people who refused to be vaccinated could be fined up to $5,000, thousands of people protested in Vienna's streets. Some of them wore symbols or carried signs comparing the mandate to the Holocaust. In Russia, where the majority of the population opposed mandatory vaccination, Moscow's mayor instituted what was then the world's largest compulsory vaccination program in June 2021. On July 7, 2022, the Chinese government mandated vaccination in Beijing. While they lifted the mandate two days after large protests, it was reported that police held people down and forcefully injected them. Eventually even those countries that did not issue a national mandate did have employers who enforced "no jab, no job" mandates. The unvaccinated were also kept out of public places to slow the spread of the disease.

Everywhere mandates were imposed, there were objections. But the best organized, largest, and most vocal protests took place in the United States. What has become clear is that in 2021 a substantial number of Americans had lost trust in both the state and federal governments. They didn't know who or what to believe.

In desperation, corporations and school boards established their own regulations. To create a safer environment for workers and customers, they began refusing service to unvaccinated people. The vax card, proof of vaccination, became a passport to participate in society. Airlines would not let people fly without proof of vaccination. Restaurants checked vax cards at the door. Schools refused to allow unvaccinated staff and students in their buildings. Businesses barred their doors, and offices insisted people work remotely. Basically, if you didn't want it, tough luck. You couldn't get into this country. You couldn't fly or in many cases go to work or school. You certainly could not work in a hospital or healthcare facility.

For some people who continued to refuse, there were severe penalties. For example, the New York City Fire Department fired firefighters for refusing to be vaccinated. A thriving business in providing counterfeit vax records sprung into existence.

I argued against forcing people to take the vaccine. These were people who had already decided they didn't want to take it for whatever reasons; forcing them only solidified their position. I believed the right strategy was spending the time and money to educate people about the advantages of the vaccine. Even if it meant having to do that an innumerable number of times, we should have taken the perspective that informed people will make the best choice. Get vaccinated; don't infect Grandma!

I did not believe it was in anyone's best interest to take away personal choice. I always have had enough respect for the American public to believe that if I can present information to them showing that being vaccinated benefits them personally and public health in general, they would make the obvious choice.

Unfortunately, that's not what was done, and as a consequence it accelerated vaccine hesitancy. Instead of understanding they were doing the smart thing by protecting themselves and their family, people took perverse pride in believing they were standing up to government oppression. That increased politicization; people weren't just deciding if they wanted to be vaccinated, they believed they were choosing between freedom and domination. It also increased polarization. Very quickly, vaccination came to be associated with the Democrats, while the anti-vax movement came to be associated with the Republicans.

A lot of anger and division might have been avoided if public health leaders had paid more attention to the data. While all strains of influenza share common properties, every strain also is different. And each of those strains can cause a different response in the body. Now add to what we didn't know about the Covid virus to what we didn't know about the value and effects of the three or more vaccines, and you end up with a steep and evolving learning curve. We made the first decisions based on what we knew at the time. As we gained experience, we tried to adapt to the new data. But at some point, that flexibility was lost, and we got locked into policies that weren't logical.

Not even when the data made it clear we were wasting precious time and resources did we change direction. Indeed, the data told us quite clearly which age-groups were most endangered by Covid. We knew that young people rarely got seriously ill, but the vaccine mandates treated all people equally. So healthy young soldiers were subject to the same restrictions as the elderly. No exceptions. People who could prove they had recently recovered from Covid already had protection, for example, and did not need to be vaccinated. But, no exceptions. Policy was policy. Rules were rules. Get the shot. Show your card.

Certainly, among the most serious mistakes was completely dismissing complaints that there might be some side effects from the vaccine. One thing that could have made a difference in vaccine acceptance and participation was simply telling the truth, even when it was uncomfortable. The anti-vaxxers successfully undermined public confidence in the vaccine with an array of attacks, but few of them made more of an impact than the claims that it caused serious side effects.

That strategy worked. People got scared, especially new parents who already were nervous about vaccinating their infants.

The "study" supposedly linking vaccines and autism had been debunked and withdrawn years earlier, but it still resonated with anxious people. The research had been fake. Others read stories on the internet about people who had a bad reaction to the Covid vaccine. But the damage was done. It is always hard to change perceptions. When people have been taught something, it's very difficult to erase it and tell them the truth. Think of Aristotle. He taught the truth. He chiseled the truth into a rock. And if it turned out to be wrong, he unchiseled the rock and started fresh again. When the government gives people information, and when it makes them take actions based on that information, it's incredibly difficult to convince people to reject what they were told is the truth and accept a new truth. It's hard for that rock to be unchiseled

There was a simple way for the government to respond to this growing distrust. Tell the truth. The data told the story. The various vaccines were incredibly safe, but in a few rare cases they caused mild side effects. Moreover, not everyone needed to be vaccinated. There were mountains of evidence to support that.

Instead, officials insisted Americans of all ages needed to be vaccinated and that there were no side effects. That just wasn't accurate. That poor guidance destroyed the credibility of public health officials. My opinion is that young children did not need to be vaccinated. The data is irrefutable. While there is nothing wrong with vaccinating kids, it doesn't provide any significant benefit. Forcing scared people to do it anyway just alienated them. Covid has almost no lasting effect on young people. The vaccine does not prevent infection. It only prevents serious illness and death, and children don't get seriously ill from this virus. They don't die from it. CDC data showed that the risk of death from Covid for people under eighteen years old was 0.01 percent. From nineteen to sixty-nine it was about 0.3 percent. The vaccine doesn't hurt them, but it doesn't help them either. The vaccine wasn't going to change those numbers. So demanding that they get vaccinated accomplished nothing positive. But it did anger parents and accelerate the erosion of trust in the public health community.

There was little science behind forcing twenty-five-year-old police officers or soldiers to be vaccinated either. But they were covered by the same regulations as elderly people who were at risk. I don't believe the vaccine should have been mandated for people under fifty years old in good health. The vaccine is not critical for those people. It was a mistake to tell them they had to be vaccinated rather than allowing them to choose to be vaccinated. Informed, intelligent people make intelligent decisions.

People were reluctant because they feared the vaccine wasn't safe. This vaccine was safe. But in some cases, people who were vaccinated did have side effects, and in a few rare cases those adverse reactions were serious. We did not have evidence of that in the first few months; but, once it became obvious that something was happening, the government should have admitted it. Would publicly discussing potential side effects cause some people not to want to get vaccinated and revaccinated? Probably. But that risk balanced against the loss of confidence in public health directives—in my opinion—is worth taking.

Throughout history, vaccines have resulted in adverse reactions. Some people who took Jenner's smallpox vaccine became feverish, or their body ached, or their arm was sore at the site of the injection. A very few people had more serious problems. That type of response has been repeated literally for centuries. People do respond to vaccines. On occasion errors are made. When the polio vaccine was distributed, several batches were later found to be contaminated, and a few even mistakenly contained live virus. The 1976 swine flu vaccine resulted in 1 person in every 100,000 coming down with potentially serious Guillain-Barré syndrome.

A major controversy erupted in the 1990s when it was claimed a preservative used in some vaccines, thimerosal, might cause neurodevelopmental disorders—including autism. While there was never any evidence supporting that accusation, just to allay fears the CDC and the American Academy of Pediatrics asked vaccine manufacturers to remove it from childhood vaccines. That was an aha moment for anti-vaxxers, who told supporters that that was proof there was something wrong with those vaccines and officials were hiding the evidence.

So the groundwork questioning the safety of vaccines had been laid for decades. The internet allowed anti-vaxxers to build on that groundwork in consequential ways. By the time Covid hit, the anti-vax movement had enormous power. The government would have benefited from being more transparent about the benefits and shortcomings of the vaccine. This had almost nothing to do with Warp Speed. As with most vaccines, safety problems only arose after the massive rollout during which millions of people were inoculated. According to the Vaccine Adverse Event Reporting System, the three most frequent side effects of the Covid vaccines were fever, fatigue, and overall discomfort. Some people also reported soreness at the injection site. No big deal; those are typical of many vaccines.

Although rare, there were more serious side effects. According to some studies, the Moderna vaccine seemed to increase the risk of swelling around the brain and heart, so several Scandinavian countries halted its use in young people. The AstraZenaca vaccine seemingly caused some blood clots, resulting in several countries suspending its use. The Pfizer vaccine was associated with an increased risk of anaphylaxis, particularly among people with a history of severe allergic reactions.

These reactions affect only a miniscule percentage of people, but they shouldn't be ignored. I remain a supporter of vaccines in general and the Covid vaccines in particular. I am so proud of the work that we did. We saved lives, tens of thousands of lives, but that doesn't mean we should ignore the dark side of it. Denying this reality contributed to the loss of trust in public health agencies.

So did giving the pharmaceutical companies immunity from prosecution. This was done to protect the companies from lawsuits that might arise from their vaccines. The companies were concerned that curtailing trials to get the vaccines out sooner could lead to increased liability. Without that legal protection, companies might not have chosen to get involved in Warp Speed. So I understand the incentive. I also understand the suspicion it provoked: Without liability, these companies were free to take risky steps. I am not—absolutely not—claiming this happened. But I still would have liked to see the Phase 4 long-term, post-marketing studies that reaffirmed the safety of their products. I believe these companies should have had the same responsibility as any other manufacturer. If at any time they became aware of safety concerns and ignored them, people who suffered should have had the right to have their injury recognized and be compensated.

It's difficult to say that the vaccine was a success when 1.2 million Americans died. But it was. It saved lives. We were able to prevent that original projection of 2.2 million deaths from becoming a reality. It also amplified an ongoing debate about the safety of vaccines in general. Today there are about twenty vaccines to fight a variety of diseases available to the American public. The public now has a lot of questions about them. Awareness is probably higher than ever before. The public should be encouraged to ask questions about safety, about frequency, about anything that prevents them from getting vaccinated.

And rather than ignoring those questions or stating simply we know better than you do so listen to us and do what we tell you, public health officials and local physicians should welcome those questions and use them as an opportunity to sell a lifesaving product. And that is done by telling the entire truth, backing it up with data when necessary. And if the data doesn't support public health policy, then those policies need to be changed.

There have been times during my career, going as far back as my AIDS discoveries, where I have been told, essentially, to shut up and go with the program. That didn't work then—that's not the way science is supposed to be conducted—and it won't work now.

Personally, I'm going to walk into my pharmacy or healthcare provider every few months, roll up my shirt sleeve, and get my Covid vaccination. I may have to wait a few minutes while Joy gets her shot. But we're going to do it, and I'm going to continue to speak out for commonsense guidance and public safety. And when asked, I'm going to continue to recommend people at risk for a bad outcome voluntarily be vaccinated.

TEN

Dr. Frankenstein's monster was not built in a day. But the real creator, author Mary Wollstonecraft Shelley, told readers how it was gathered piece by piece and reassembled. It would be extremely beneficial to know where the SARS-CoV-2 virus we fought came from. Knowing its origin might allow us to take steps to prevent a tragedy like this from ever arising again. The Frankenstein story ends with the monster adrift on an ice floe, disappearing into a self-imposed exile. Unfortunately, Covid is here to stay.

When the virus first appeared in the first few days of January 2020, before we had a single case here, Tony Fauci and I met to discuss what we knew and what to do about it. It wasn't much, just that a potentially deadly pathogen was circulating in Wuhan, China. Even then, at the beginning, we wondered: Where did it come from?

There were two hypotheses. First, it was created naturally and emerged from the Wuhan live animal and seafood market, jumping from species to species, until it finally figured out how to infect human cells. Second, it had been created by scientists conducting gain-of-function experiments—what would happen if we did this?—at the Wuhan Virology Institute, and it had escaped from that laboratory. As a clinical virologist I believed that the origin of Covid was a direct consequence of science, in particular gain-of-function research. Fauci strongly disagreed.

At that time, it was impossible to reach any conclusion. The debate continued without any firm resolution. That is not surprising. To this day, researchers still have not been able to firmly identify the cause of the 1918 pandemic. While there is evidence to support both theories—the market or the lab, nature or humankind—I have made up my mind.

Life is the greatest scientific mystery. Scientists have spent centuries slowly unravelling its secrets. So it certainly isn't surprising we don't know how viruses were created. But we do know they have plagued human existence through history. Evidence found in neanderthal skeletons, permafrost, and artifacts indicate that viruses—among them herpes, hepatitis B, and papillomavirus—existed 50,000 years ago. Indications of smallpox and polio have been found in Egyptian mummies dating back to 1500 BC. There is no question that viruses are created by nature, perhaps from clashes of proteins from different species or a mix of DNA and RNA.

We can't create viruses in a laboratory. But we can change them.

Figuring out where the Covid virus came from, especially if it was modified in a laboratory, would enable us to make it less likely it'll recur. Initially, we believed that Covid was a variation of SARS and MERS. Both of those viruses apparently originated in bats and then were moved adaption by adaption through animals into humans—SARS through civet cats, MERS through camels—so it was natural to assume Covid had followed a similar path.

But neither the SARS nor MERS virus learned how to effectively transmit from human to human. That's why they caused only a limited number of cases. Covid was very different. Covid had no problem infecting hundreds of thousands of people. It is one of the most infectious viruses we have ever encountered. Why? What makes it so different?

Then we got the first puzzling clue. Apparently, the COVID-19 virus cannot infect bats. Somehow, if it did come from bats, it must have changed so significantly in the transmission process that it lost the ability to bind to bat cells. Researchers have spent the ensuing years searching for the animal that served as the go-between allowing it to go from bats to humans. No one has found it. For me, that argues against the natural creation theory.

I believe the virus was engineered to infect human beings.

It was taught how to do it. I also believe it was made by researchers in a laboratory at the Human Virology Center in Wuhan and somehow escaped to change the world.

It also is important to add that I do not believe this was intentional. To be sure, there are potentially sinister shadows. The lab in Wuhan does have a military component, just as we have our virological research center at Fort Detrick. But it makes absolutely no sense that the military would create and release a virus when they have no way of controlling it. The first victims of such a strategy would be the Chinese people. So there just is no reasonable motive nor logic in that scenario. In fact, from the people I know in China, from the discussions we've had and everything I've read, my educated guess is that it came from the nonmilitary side.

No matter how secure the precautions are, viruses can escape from laboratories. It happens. Even with every possible precaution taken, there have been incidents. In the 1950s, for example, two men working at Fort Detrick died from exposure to anthrax. A decade later, a worker died from viral encephalitis.

The American government currently operates more than two hundred biological laboratories in twenty-five countries and regions around the world. According to *USA Today*, there were more than a thousand incidents involving viruses, bacteria, and toxins that could have posed serious danger to people and agriculture between 2008 and 2012. Since then, there has been an average of two hundred events involving these agents reported each year.

Fort Detrick is probably the largest and most secure of our biological laboratories. The American military began conducting research and some biological experiments there during World War II. In 1969, President Nixon, after signing the 1925 Geneva Protocol prohibiting the use of chemical and biological weapons, prohibited all research to create new biological weapons. Since then, Fort Detrick has remained at the center of efforts to defend America from biological warfare. It is a large, heavily secured military post containing some of the most sophisticated biological laboratories in the world. It is the type of top secret facility novelists write about, and much of the research and development that has taken place there to help protect this country will remain confidential. If the walls could talk, they would not be permitted to.

During my military career, I spent considerable time there. I knew the place well. The CDC has the responsibility of monitoring high-containment labs to ensure they remain safe and very secure. During a 2019 inspection, we found six serious violations, including two containment breaches. There were leaks and mechanical problems with a recently installed chemical system to decontaminate wastewater. These were reported as failures to "implement and maintain containment procedures sufficient to contain select agents or toxins."

Several weeks before that inspection, we received reports that a mysterious respiratory illness had broken out in the Greenspring Retirement Community in Fairfax County, Virginia, that had resulted in sixty-three cases and three deaths. Greenspring is about an hour from Fort Detrick, close enough for us to pay serious attention.

I ordered Fort Detrick shut down. I knew what could happen if a pathogen escaped from a biological lab into the community. There were people who criticized me for that, thinking I was acting too conservatively. I didn't think I had a choice. That wasn't going to happen on my watch. We kept it shut down for seven months.

Biological laboratories use a variety of physical barriers to contain pathogens. These include sealed secure doors, double doors, airlocks, and physical separation from other sections. The flow of air going in or out of the lab is filtered. Access is limited, and every person working inside has to wear full personal protective equipment. They leave all that equipment inside the lab. In the higher levels of containment, they shower before exiting. Still, pathogens escape. In 1977, a version of the H1N1 virus, a modified version of the pathogen that caused the 1918 pandemic, escaped from a lab in China and triggered a massive effort to immunize Americans. Between 1966 and 1978, there were three incidents of the smallpox virus escaping from labs in Great Britain. In 1979, anthrax bacteria escaped from a bioweapons lab in Sverdlovsk, then the USSR, and ended up killing at least sixty-four people. In 2004, there were two separate incidents of SARS somehow escaping from a Beijing lab. In 2007, 75,000 Venezuelans were infected and three hundred people died when equine encephalitis got loose from a lab there. Another UK incident in 2007 sparked an outbreak of foot-and-mouth disease.

When I learned a virus was spreading in Wuhan, I couldn't help but wonder if it had escaped from the Institute of Virology there. I had visited that country several times. In 2000, the Chinese CDC invited me to advise them on the AIDS epidemic then beginning to spread there. They shared all their data with me, they allowed me to travel to all the different parts of the country, and that's when I developed my relationship with George Gao and other leading scientists. While there, I had also visited several wet markets. They were crowded places of organized chaos. So I didn't dismiss the possibility it had started in a wet market. To be honest, when the virus first began spreading in Wuhan, I simply did not have enough evidence to reach a conclusion.

Much of the work taking place in biological labs remains secret. Few outsiders know what research is being conducted. But among the most controversial and challenging work is something called gain-of-function research. That term itself is relatively recent because the technology itself only recently became possible. Very basically, it means genetically changing microbes—such as bacteria and viruses—to give them properties that do not come from nature. What would happen if I moved this little protein over there? Our primary concern is any change in a microorganism's ability to interact with human beings. It often means making them more transmissible or more virulent: easier to infect, more dire consequences. Until the pandemic, few people outside the scientific community were even aware this type of research was being conducted. Gain-of-function research can be beneficial. In some ways, we're creating new worlds, where possible treasures will be found. This research can lead to the discovery of a defense against a dangerous pathogen. It might be the basis of a vaccine. It can help provide clues to biological puzzles. But in new uncharted worlds, there are also dangers.

We knew from the beginning this virus had come from Wuhan, China. But where in that region and how? Was it natural or was it created by scientists? Did it rise out of the wet market from animals or from research gone awry? And more ominously, did the United States at least partially fund that research?

Even with all the evidence, the official Chinese position is that Covid did not start there. The Chinese Embassy to the United States issued an indignant statement stating flatly, "For some time, various lies and rumors concocted by the US side against China on origins tracing have been repeatedly refuted by China and the international community with detailed facts and data. . . . Origins tracing is a scientific matter, yet the US side is using intelligence agencies to trace the origins, and peddling the old lies that have been refuted under a cloak of intelligence. Washington's real purpose is attempting to confuse the public and deceive the world, and continuing to seek a 'presumption of guilt' against China, politicize the origins tracing, shift the blame onto China, and suppress and contain it."

That part of this statement, "origins tracing is a scientific matter," is absolutely accurate. And there is an abundance of good science proving beyond any doubt that the virus originated in China.

Suspicion immediately was raised that it had been created by the military in the Virology Center there. Naturally, social media picked up that thread and magnified the speculation into a rumor. The president began referring to it as "the China virus." That was accurate, but it wasn't especially helpful. The entire world was facing a massive threat, and it needed to unite to save lives. I considered it inflammatory and counterproductive to target the Chinese for the pandemic. There's no benefit in putting labels like that on it. The Spanish flu, for example, probably came from the United States. The disease became known by that name primarily because Spain remained neutral during World War I; unlike the press warring nations, whose censored newspapers did not report the growing death toll, the Spanish papers wrote about it, creating the impression it had started there. The Spanish people were irate at being unfairly blamed for it. Calling it the Spanish flu provided no benefit while alienating citizens of that country.

I knew that we had a core group of dedicated scientists within China working their hearts out to understand this. It would have been more fruitful if the United States had taken the position that the entire world is facing a new pathogen, let's get together and figure out how to confront it scientifically.

I spoke often with my Chinese counterpart, George Gao. I never doubted his sincerity. He was very honest with me about the outbreak there being out of control, that large numbers of people were being quarantined. But there also was no doubt about the political involvement. The tone of our conversations changed as we moved into April. He became less forthcoming; there was more hesitation in his voice. He would pause thoughtfully before speaking, making our conversations more formal and less interactive. There is an odd feeling when you're talking to a friend that something has changed. It began to feel like we were communicating through a wall. In his defense, our calls were being recorded by the Chinese government. No surprise, just like our NSA taps into calls.

I never found out how much or what George Gao knew about where this came from. If he did have that information, I never got a hint of it. In China, there are four independent levels of public health agencies. Like here, access to information can be much less transparent than we would like to believe. In 2024, Gao told an interviewer in London he personally had not reached any conclusion. "You can always suspect anything," he said. "That's science. Don't rule out anything." But he also questioned the wet market theory. "Even now," he added, "people think some animals are the host or reservoir. Cut a long story short, there is no evidence which animals (were) where the virus comes (from)."

I know George tried and, unfortunately, failed to stop the virus. When that failure became evident, he tried to control it, and he failed at that too. No one knows how many people ultimately died in China. I'm sure it's exponentially higher than we believe. They died because a deadly pathogen once released into the world cannot be controlled. You release an agent into the environment, and, as a result, you can't occupy that environment? It makes no sense. The Chinese know that. That's why I do not believe this was intentional. Not unless they intended to cause a massive health and economic crisis within their own country.

But conspiracy theories continue to flourish as the world looks to place blame. In fact, Chinese government officials and their national media launched a campaign suggesting the SARS-CoV-2 virus had escaped from . . . Fort Detrick. "The COVID epidemic hit America in April 2020, and New York became the epicenter," according to Chinese social media website WeChat. "In the meantime, at Fort Detrick about 240 miles away, the US government was conducting experiments with dangerous pathogens." It also was noted in the Chinese media that the facility had been shut down only several months earlier—due to contamination.

The Chinese also suggested the virus might have unknowingly been brought into that country in the frozen carcass of imported beef from an unidentified nation.

Unlike the Chinese claim that it had come from America, a reasonable argument can be made to support the natural theory. There are researchers who believe Covid was not created in the wet market. Instead, they believe it came from Western China or Northern Laos, before it was carried into the Wuhan market by horseshoe bats or animal trade. They base that on research showing that three strains of viruses found in Laotian bats are genetically closer to SARS-CoV-2 than any previously known viruses. These strains can be tracked along the same path that resulted in the 2002 SARS epidemic, ending up in already infected animals being brought to the wet market.

Obviously, none of that is true, but it is possible to build a reasonable circumstantial case using mostly facts. That's the problem with determining the origin. There are more questions than answers. One thing on which there is general agreement is that this virus came from horseshoe bats.

There are only two possibilities for how it leaped into human beings: natural or manmade. There has been a tremendous amount of investigation into this question, and the result is a lot of opinion. The World Health Organization concluded in 2021 that this was likely a "zoonosis event," meaning it made the leap from animals to human. The WHO added it is highly unlikely it was caused by a lab leak. Conversely, after two years of hearings and research, a House Committee issued a massive report speculating Covid most likely leaked from the Wuhan lab. They cited several broad reasons. For example, "the virus possesses a biological characteristic that is not found in nature." And "data shows that all COVID-19 cases stem from a single introduction into humans. This runs contrary to previous pandemics where there were multiple spillover events."

No matter how it got there, there is general agreement it first surfaced in that market. Most of the first cases could be associated with people who worked at or visited that market. And after the government shut it down, researchers found the Covid virus had spread widely there, particularly in the section of the market where live and freshly butchered animals were sold. Researchers were able to identify the stall where one of the positive samples had been found. A photograph of that stall taken several weeks before the outbreak shows a raccoon dog, a popular Asian breed related to a fox but resembling a raccoon, sitting there; a dog's DNA was found mixed in a sample of the virus. In a truly great stretch, there are people who wonder if the virus was created in that stall.

Among the people who favor the wet market theory—with limitations—is Tony Fauci. "The environmental swabs that were taken from that part of the market, where there's photographic evidence that there were raccoon dogs, again is accumulating evidence that it was a natural occurrence," he said, but then added, "you have to keep an open mind until you definitively prove one versus the other.

"We may never know for certain. It took years and years to find the original source of HIV in chimpanzees. It took several years in SARS-Cov-1, which came about in 2002 or 2003 to show that correlation between a bat to a civit cat to a human. We still don't know the primary origin of Ebola."

Of course, there is another possibility that I believe is far more likely: It is probable that one or more already infected human beings brought the virus into the market, spreading it to animals and humans. The question becomes where did they get it? Many people continue to believe it occurred naturally. I'm not among them.

While I have supported aggressively investigating both theories, the data makes me believe COVID-19 infections were the direct result of biomedical research and a subsequent lab leak. There are several reasons to believe it came from the Virology Center. There is extensive evidence that cutting-edge research about these viruses was taking place there. One of the leading scientists there was Dr. Shi Zhengli, the director of the Center for Emerging Infectious Diseases at Wuhan. Dr. Shi is a well-respected researcher and had said repeatedly that the virus did not come out of her lab. There is no evidence that the virus was created there or escaped from there. So this is not an accusation.

But the coincidences are compelling. For two decades Dr. Shi Has been focusing on coronaviruses that come from bats. More than two decades ago, she discovered that bats are efficient carriers of SARS-like coronaviruses and since then has made advances in our understanding of this complex relationship. Her teams have traveled throughout China collecting thousands of virus samples, enabling her lab to compile one of the largest bat-related virus databases in the world and earning her the title, China's "batwoman." She has been a leader in studies of the receptors in human, animal, and bat cells trying to figure out how viruses made the leap between species.

It also is known that she has collaborated in genetic engineering experiments with American researchers with both SARS and MERS coronaviruses. A paper she published in 2017 likely identified the source of the SARS outbreak as horseshoe bats found in caves in China's Yunnan province. But another paper she published that year described an experiment in which her lab at the Wuhan Institute mixed the DNA of several bat coronaviruses, among them to create a new virus. There was nothing nefarious about that. The objective was to better understand how these viruses can be transmitted to humans to cause terrible cellular damage and to find a way to prevent that from happening. A lot of people other than myself were certain a pandemic was coming, and there has been a worldwide race taking place to develop a way of protecting us from these viruses. As Dr. Shi later explained, "We must find them before they find us."

While that might sound like a reasonable justification for conducting that research, in fact I have always been strongly against it. It is my belief that the potential danger greatly outweighs any benefit. Just because we can do it is not a reason to do it.

We lost that round.

In March 2019, only months before the first record Covid cases, she and three colleagues published an article in the respected journal *Viruses* citing the fact that there had been three large-scale disease outbreaks caused by coronaviruses spilled over from animals with common characteristics. "They are all highly pathogenic to humans or livestock, their agents originated from bats, and two of them originated in China. Thus, it is highly likely that future SARS- or MERS-like coronavirus outbreaks will originate from bats, and there is an increased probability that this will occur in China."

As Covid began spreading, Chinese social media and American reporters speculated on the possibility her lab was involved. Dr. Shi vehemently denied every accusation, refuting rumors that several of her colleagues may have been ill at the beginning of the outbreak. She angrily told reporters, "How on earth can I offer up evidence for something where there is no evidence?" In a text message sent later she added, "I don't know how the world has come to this, constantly pouring filth on an innocent scientist."

American intelligence agencies named three scientists at the Institute as potentially being patient zero, the first known person to be infected with Covid. One of them, Ben Hu, called that report ridiculous and dangerous, saying he had never been ill in the weeks before the disease became reality.

While Dr. Shi is recognized as a leader in research, there are many other labs around the world—in China, the United States, Japan, Hong Kong, Singapore, Australia, the UK, and other regions—conducting similar research. Among the people who collaborated with her was Dr. Ralph Baric, an award-winning virologist at the University of North Carolina. Dr. Baric has been researching coronaviruses for decades and has published numerous reports. A 2015 paper he coauthored showed that coronaviruses carried by bats could pass directly to human beings rather than requiring an animal to transit it. That made sense to me. In my opinion, I just don't think it's plausible that this virus went from a bat to another animal—we still don't know what animal—before going to humans and immediately learning how to be human-to-human transmissible to the point of now causing one of the greatest pandemics we've had in the history of the world.

If, as I believe, it was created and escaped from a lab, the next obvious question is: Why was it created? My answer is scientific arrogance. Just because we can do something is not enough reason to do it. It is possible to be too smart for our own safety. In about 2010, scientists in the Netherlands genetically altered a potentially deadly strain of bird flu to make it more easily infectious. And they did it intentionally. That's arrogance. H5N1 was known to kill about 60 percent of the people it infected, but fortunately it required very close physical contact between a bird and a person to spread this virus. But scientists at Erasmus Medical Centre in Rotterdam and the University of Wisconsin took the virus through a series of mutations essentially to show how simple it is to change its genetic properties, turning it into a highly infectious airborne strain that could be spread by sneezing and droplets. Fortunately, these changes also made it considerably less lethal. These experiments were conducted on ferrets, which are commonly used as substitutes for humans in influenza research. According to published reports, these experiments were at least partially funded by the National Institute of Allergy and Infectious Diseases, which was then directed by Tony Fauci.

Their reason for conducting this gain-of-function research, they explained, was to help scientists create new vaccines and drugs. I argued vehemently that researchers shouldn't be doing this research until we can be certain there is no chance of leakage. In other words, never.

I feel strongly there is no scientific advantage in conducting this type of research, and the obvious downside is that we are creating pathogens that may cause great harm, for which we don't have countermeasures—like the COVID-19 virus, which I do believe was a direct consequence of gain-of-function research.

Many of my colleagues disagree with me. In an opinion piece for the *Washington Post* in 2011, Fauci, Francis Collins, and Gary Nabel, director of the Vaccine Research Center, defended gain-of-function research, writing, "research has allowed identification of genetic pathways by which such a virus could better adapt to transmission among people. This laboratory virus does not exist in nature. There is, however, considerable concern that such a virus could evolve naturally. We cannot predict whether it or something similar will arise naturally. . . . Given these uncertainties, important information and insights can come from generating a potentially dangerous virus in the laboratory. . . . Determining the molecular Achilles' heel of these viruses can allow scientists to identify novel antiviral drug targets that could be used to prevent infection in those at risk or to better treat those who become infected."

They concluded this piece by acknowledging—and accepting—the risks. "Safeguarding against the potential accidental release or deliberate misuse of laboratory pathogens is imperative. The engineered viruses developed in the ferret experiments are maintained in high-security laboratories. The scientists, journal editors and funding agencies involved are working together to ensure that access to specific information that could be used to create dangerous pathogens is limited to those with an established and legitimate need to know."

I understand the argument. I do. I just don't agree with it. Some proponents suggest we have to create these dangerous viruses to learn how to develop countermeasures to them. I don't believe that. And I have yet to see any evidence that this research has resulted in life saving benefits. We have enormous scientific resources so if and when we are challenged with a new pathogen, we can apply the tools of science to defend society against those pathogens.

Nature has created an extraordinarily efficient balance. Humankind grew up in harmony with our environment. We are the result of learning how to live in that environment. Many of those things that could kill humans were dealt with—either eliminated or controlled. Sometimes it required weapons, but often we were able to use science. Then, as humanity expanded into different environments, we maintained that balance. We used the knowledge we had acquired to figure out how to deal with new challenges, new threats. We found remedies and preventions. In some situations, we began making modifications to nature in laboratories—many of them productive and beneficial. We learned how to control or in some cases prevent many deadly or crippling diseases. We figured out how to modify the human systems on a cellular level. Things that once killed us were transformed into minor nuisances. People gained confidence in the ability of scientists to keep them safe. As the poets write, we became masters of our fate.

Or so we have come to believe.

Doing it was questionable enough, but then they decided to publish the details of this experiment. Basically, this would hand the recipe for creating a bioweapon to terrorists. There was a long, controversial debate that resonated throughout the entire scientific community about whether or not this type of experiment should be conducted and if so, should the results be published?

Science progresses in very small steps. That happens because scientists share data. Not all information is shared, not the secret formula for soft drinks for example, but the scientific method literally requires discoveries be published so other people, working independently, can prove or disprove them by attempting to reproduce those results. Sir Isaac Newton wrote famously to a colleague in 1675, "If I have seen further, it is by standing on the shoulders of Giants." His meaning was clear: Only by building on the work of others could he have made his discoveries. Publishing allows claims to be peer-reviewed, to be tested to determine their accuracy.

There is considerable pressure on researchers to publish. Being published in respectable journals brings attention which can benefit a career or lead to grants. As I learned when I published several papers about my work on AIDS, that information also can create controversy. It can cause a lot of debate, sometimes unpleasant debate.

But there also has been general agreement that some things should not be published.

For many people, this was one of them. Lynn Enquist, editor-in-chief of the *Journal of Virology* and a member of the National Science Advisory Board for Biosecurity—a committee of expert virologists that provides recommendations to the government on potentially dangerous life science issues—remembers, "We had never seen any papers like this with the gain-of-function idea. The idea of increasing virulence, or increasing transmissibility, was not really something that most scientists had ever thought about doing. It was a concern."

That board suggested the conclusions be published, but not the details of the experiment nor the data. As Paul Keim, the chairman of the NIH's biosecurity board, warned at the time, "We were saying, 'Wow—it's highly transmissible with a 60 percent mortality rate,' You could kill 4 billion people in a flash, because these viruses go around the world." They also urged a moratorium on this type of research, and in response those labs agreed to a brief, temporary pause. After about two years of debate, these experiments were published, in full detail; one of them in *Nature*, the other in *Science*. Both respected journals.

Ron A. M. Fouchier, who conducted the experiments in the Netherlands, defended his decision to publish, writing in *The Journal of Infectious Diseases*, "The likelihood of the airborne A/H5N1 virus being used by individuals or organizations with bad intentions is low. The possibility to do harm with this virus is probably low in terms of its transmission and virulence in humans, because it is unlikely that the virus would spread like a seasonal influenza virus in humans and because the case-fatality rate is likely much lower than currently estimated. The techniques that we used to create airborne A/H5N1 virus are not new and can be found in many virology textbooks. Individuals with bad intentions do not need to read the details in our manuscript because the methods for creating similar viruses have already been published widely." For Fouchier, censoring the manuscripts would "only create a false sense of security.... The more danger a pathogen poses, the more important it is to study it."

The likelihood of doing harm with this virus is "*probably* low"? When millions of lives are on the line, that isn't good enough. The potential exists that our enemies can use this information to create a virus capable of killing millions of people. So "probably low" definitely is not good enough. There is a huge difference between studying a pathogen to learn its capabilities and changing its properties. The fact that this was done in a university lab rather than a secure government facility makes it even more troubling. I can say with some confidence when this type of research is done in a university lab—which is not at all uncommon—we can anticipate these viruses are going to escape.

The wisdom of publishing this material continues to elude me. If the goal is to make it easier for other people to conduct similar research I have to ask, again, why? There is little benefit and large danger in making it easier for other—perhaps less qualified—researchers to do this. And there are always bad actors out there who might be able to make use of this road map.

It is my belief that those centuries of scientific success have made us woefully overconfident. Or, as I describe it, arrogant. We continue to believe we can control whatever we create. We literally believe we can manipulate nature in laboratories and there won't be any consequences.

That's a false belief. We know for a fact we can capture potentially deadly pathogens and try to uncover their secrets. We know for a fact we can change the properties of those pathogens to make them less deadly—or more deadly. But we also know for a fact we can't keep our laboratories secure.

Gain-of-function research is a recipe for disaster. And knowing all that is why I believe COVID-19 was created in a laboratory.

Let me be clear about this: Admittedly, none of the information that has been published demonstrates in any way that COVID-19 was created in a laboratory and leaked into society. There are other reasons I have reached that conclusion.

As I have explained, in order for a virus to infect a cell it has to attach itself to that cell wall. It has to find a place to dock, a receptor site. If the virus can't find a receptor, it bounces off, which is what happens with almost all bat viruses—among them SARS-CoV-2. Until one day it doesn't bounce off.

In 2020, I was asked to meet with Secretary of State Mike Pompeo. In the previous months, I had become an unofficial advisor to him. My job was to interpret science for him. At the beginning of this meeting, he handed me a highly classified intelligence document. "Tell me what this means, please," he asked. He sat silently for several minutes as I read it.

I was stunned. It isn't often such clear-cut scientific evidence is just handed to you, but this report laid out the reason this formerly benign virus had suddenly become a killer. The report has since been declassified. "What do you think?" he asked me.

I told him, "Mr. Secretary, this is the smoking gun. It confirms that this virus did not come from nature. That it was engineered in a lab to infect human tissue."

I explained it to him. Investigators had discovered the existence of something called the furin cleavage. Basically, it is a unique combination of twelve nucleotides that have been arranged to create the amino acid known as the furin cleavage site, the receptor that welcomed this virus. The virus can't infect humans without the furin cleavage site. If you remove the furin cleavage site from the COVID-19 virus, there is no pandemic. There's no human infection. 20 million people around the world would still be alive.

Epithelial cells in bats contained a receptor allowing this virus to bind to it. But when the furin cleavage was placed in COVID-19, it changed the orientation of that binding site in bats, so the virus no longer recognized the bat receptor. Somehow the virus that initially had been carried by bats had been changed so it no longer could infect bats. But once again, if you remove the furin cleavage site it sticks to the bat receptor like glue and bounces off the human receptor.

The furin cleavage site was well known to researchers. It was first recognized in a lab in the 1960s. By 1972, it was found in certain viruses, but nothing that later would be related to coronaviruses. As Nobel Prize laureate David Baltimore explained in 2021, "Within the SARS-Co V-2 genome there is an insertion of 12 nucleotides that is entirely foreign to (this strain) of virus that SARS-Co V-2 is in. There are many other viruses in this class, including the closest relative of SARS-Co V-2 by sequence and none of them have this (furin cleavage) site."

At first Baltimore agreed that this was "the smoking gun." He has since backed off a bit. "There are other viruses that have furin cleavage sites, other coronaviruses, though not the family of beta-coronaviruses. So this sequence's nucleotides could have hopped from some other virus. No one has identified a virus that has exactly this sequence, but it could have come from something close, then evolved into the sequence that we see today. . . . I'm perfectly willing to believe that happened, but I don't think it's the only way that that sequence could have appeared. The other way is that somebody could have put it in there. You can't distinguish between the two origins from just looking at the sequence. . . . When I first saw the sequence of the furin cleavage site—as I've said other beta coronaviruses don't have that site—it seemed to me a reasonable hypothesis that somebody had put it in there. Now, I don't know if that's true or not, but I do know that's a hypothesis that must be taken seriously."

Since then, the search has continued—without success—to find this unique cleavage in the same strain of virus. Still, many respected researchers continue to believe this was a naturally occurring mutation. When you look at it as an isolated event, that argument is as valid as anything else. But it wasn't isolated. In order to believe that it was a natural event, you have to pretty much ignore the gain-of-function research then being conducted at Wuhan with this virus and the questionable security of those labs. All of that, combined with the fact that no one has been able to identify the animal that might have served as the conduit, makes me confident it came from a lab.

And I know what that portends for the future.

As it turns out, a majority of people seem to agree with me. As did Secretary Pompeo, who told reporters, "I can tell you that there is a significant amount of evidence that this came from that laboratory in Wuhan."

At that point, opinion was decidedly mixed. Less than a third of Americans believed it came out of a lab, Since then, that has changed significantly. By 2025, about two-thirds of Americans agreed with me that Covid was engineered by researchers. Around the world, opinion continues to move in that direction. In 2025,

German intelligence agencies reported with significant confidence that it leaked from a lab. The French Academy of Medicine almost unanimously supports the belief it leaked from a lab.

As investigations continued into how this happened, surprising information emerged, including the probability that the NIH not only supported gain-of-function research being done at the Wuhan lab, it paid for it—at least partially. There is absolutely no evidence that any of this was ill intentioned. It certainly wasn't an act of bioterrorism. But it was misguided. The intent was to understand what permits this virus to replicate more efficiently and prepare for it. It's like building an army for a war that might never come. It's a defensive strategy.

When this claim that the NIH funded gain-of-function research at the Wuhan lab first arose, Tony Fauci was adamant that did not happen. At one point, he told a congressional committee, "The NIH has not ever and does not now fund gain-of-function research in the Wuhan Institute." The debate became enmeshed in political maneuvering. It basically has come down to an interpretation of the meaning of gain-of-function.

But in May 2025, Dr. Jay Bhattacharya, the NIH's new director, admitted during a staff town hall, "It's possible that the pandemic was caused by research conducted by human beings. And it's also possible that the NIH partially sponsored that research."

He went on, "If it's true that we sponsored research that caused a pandemic—and if you look at polls of the American people, that's what most people believe, and I looked at the scientific evidence, I believe it is what we have to do is make sure that we do not engage in research that's any risk of posing any risk to human populations."

Obviously, I agree with him. Any sensible person does. We may never find definitive evidence proving one way or another the origin of the virus. But that should not stop us from heeding Dr. Bhattacharya's warning. The fact is that research currently being conducted in insecure labs around the world has the potential of causing the next pandemic.

And as a result, millions of people may die. Millions and millions and millions . . .

ELEVEN

Use whatever metaphor you'd like: The clock is ticking. Darkness is coming. It doesn't matter, it will all have the same result. The next pandemic is coming.

Looking at history, we've had influenza outbreaks in the United States about every twenty years. You can map them out from the 1800s. There is no reason to believe that will change. If anything, it will accelerate. The fuel for that might well be climate change. I've always advocated studying the health consequences of climate change. We have already seen significant changes in pathogens because of temperature change. Now add to that the mass migration of living things in response to the impact of climate change; people, animals, birds, rodents, and even ticks and fleas are changing environments. As a result, more people are living in closer contact with limited hygiene than perhaps ever before in history. That's pretty much the textbook definition of a breeding ground for disease. In the world of infections, proximity facilitates transmission.

The other side of that equation is the adaptability of viruses. They evolve to survive, and they are very good at it. Their ability to massively and rapidly reproduce until they discover a path to infection makes it difficult to mount a defense. Viruses change, and change, and change. And once created, they do not go away. Instead, they create variants. That's why we produce a new flu vaccine every season.

Borders are not barriers. It isn't possible to prevent the spread of a virus. The predictable migration patterns of birds, for example, tracks roughly with viral outbreaks. Why, for example, do we suddenly see a sudden outbreak in a Nigerian community? Years ago, while I was working in China, they showed me extensive studies in which the spread of the flu essentially mirrored the annual bird migration—right though that Nigerian city.

Historically, animals, rodents, birds, bats, mosquitoes, ticks—and labs—have been responsible for igniting epidemics. Having studied and combatted viruses for decades, I strongly believe the next pandemic will come from bats, birds, or beakers.

SARS, MERS, and COVID-19 all came from bats. What is unusual about these viruses is that normally when a new pathogen infects human beings it's frequently characterized by high mortality. It doesn't come into the human species and say, "I'm a gentle pathogen, I'm not going to cause a lot of problems. We can live happily ever after." Usually they act like HIV—they kill the host. Covid was different; about half the people it infected showed no symptoms whatsoever.

Bats are very efficient at spreading disease. They are the only flying mammals, they have a relatively long lifespan, they gather and roost in large colonies—often in dank caves, which facilitates spreading—and they have a unique immune system that enables them to be asymptomatically infected. For some reason, we don't quite understand why they can carry viruses fatal to other species without showing any effects.

Bats have played a role in mythology for centuries. Da Vinci's futuristic flying machine drawings with broad wings might well have been based on bats. Bram Stoker's *Dracula* was first published in 1897; Dracula's ability to transform into a bat played on the sinister impression that bats lived in the dark and infected humans. Even the 1939 comic book character Batman was created because a vindictive Bruce Wayne wanted a disguise that would "strike terror into the hearts of criminals. I must be a creature of the night, black, terrible."

The fact is people know more about Batman than bats. Until recently, research into bats was somewhat limited. Few people know, for example, that there are more than 1,400 species of bats. Part of the reason for that, I assume, is that interactions between people and bats have been limited. Bats traditionally nest in places not frequented by humans. But as human populations have expanded into new environments, that has changed. Disease-carrying bats have become a greater threat than ever before. To prepare to defend against that threat, researchers around the world are now studying bat viruses in labs. Ironically, we have seen, that increases the threat.

It is more likely though that the next pandemic will originate in birds. Birds meaning every type from the small songbirds that stop briefly in our yards to the vast poultry industry. It's difficult to avoid any contact with birds. And the disease they carry—bird flu, also known as H5N1 or HN79, and its variants—is here and spreading. It was first identified in 1996, but for several decades was confined to domesticated birds. But eventually it spread to wild birds. For several decades, it didn't really affect infected birds. They didn't get sick. The only way to determine if a bird was infected was by testing it. Gradually, that changed. It began killing birds. It also began to pose a significant danger to anyone who had contact with those birds.

H5N1 is highly infectious in birds. It is spread by direct contact, contaminated equipment, bird droppings, through the air, and by human beings carrying it on their clothing or tools. There is no treatment. It has been found in about four hundred bird species and has spread to at least fifty mammals. It is in bears and bobcats, in dairy cows and domestic cats, in pigs and dolphins, and it is continuing its spread. Fortunately, so far, it hasn't learned how to go from mammal to mammal; thus far its transmission method has been bird to mammal, bird to mammal.

But it is evolving.

In 2023, H5N1 essentially wiped out Argentina's entire population of Southern elephant seal pups. Somehow a single seal was infected, probably from contact with a bird, and the virus mutated, enabling it to spread wildly. Seventeen thousand carcasses washed up on Argentina's breeding beaches, strong evidence that mammal-to-mammal transmission is possible. That same year, the virus killed a polar bear in the Arctic. In other words, H5N1 is not deterred by frigid temperatures, and it has now spread to the most distant regions of the earth.

There have been occasional human infections. In 2024, the virus began infecting dairy cattle in the American Southwest, eventually spreading to more than one thousand herds in seventeen states and infecting dozens of farm workers. While the farmers' symptoms were mild, that generally is not the case. This is a deadly disease. The World Health Organization has confirmed 972 bird flu infections in human beings since 2003, leading to 468 deaths. It has killed just about half of all the people it has infected. By contrast, Covid has killed less than 1 percent of all people infected.

In January 2025, the first person in America died from bird flu. This was an elderly Louisiana man with an unspecified underlying medical condition, who was infected either by the flock of chickens he kept in his backyard or a wild bird. He was hospitalized for several weeks before succumbing to the disease. After his death, the CDC sequenced the virus. They discovered that the strain that had initially infected him had changed in his body. It had learned how to replicate better in humans.

Better. Fortunately, still not easily. H5N1 does not know how to bind well to mammals. It might inadvertently infect an animal or human being—usually from direct contact with birds, including saliva and droppings—but it doesn't know how to take hold, replicate within that system, and pass from mammal to mammal. *Yet.* Each time it infects someone it keeps trying and keeps changing. There are several different strains of the virus in circulation. In an effort to contain it, more than 165 million birds, including chickens, turkeys, ducks, and other poultry, have been slaughtered since 2022, leading to a brief increase in egg prices. But there is no way to contain it in wild birds. Or in laboratory experiments.

My biggest fear is that it may have human help.

I can take the virus—or any one of the numerous other pathogens—into a laboratory and within a week teach it how to infect human beings. Just as I believe researchers in Wuhan did with Covid. This is the result of publishing the details of those experiments more than a decade ago. We know the structure of the amino acids in the bird flu virus, and we know exactly how to teach them to find a receptor on human cells.

I believe that biosecurity is the most dangerous national security threat facing this country. There is more danger to our way of life coming from nature and labs than from munitions plants. Covid made the entire world pause. It killed millions of people, it cost trillions of dollars in economic activity, and it exacerbated existing political divisions both in this country and internationally. But that is minor compared to the potential of the next pathogen that mutates in nature or is enhanced in a lab. It may be the result of a gain-of-function accident. Or biological warfare. Or bioterrorism.

This country has enemies. Smart enemies capable of creating havoc. China, Russia, and North Korea, for example, all have sophisticated scientists working in labs to create . . . we don't know. It would be a pleasant surprise if they were not conducting gain-of-function experiments. In fact, it doesn't even have to be a state supported program. It does not require tens of millions of dollars to conduct this research. With the information publicly available, experienced scientists working for terrorists or a criminal organization might be able to produce an enhanced pathogen. If they succeeded—and were prepared to sacrifice themselves—our national security would be in jeopardy.

How do we prevent this? I'm not sure we can, but there are steps we can take to decrease both the probability that it will happen and the destruction to life and society it will cause. Certainly, we have to increase lab security. I was surprised how lax procedures were when I became CDC director. We immediately took steps to examine the existing standards, harden them where necessary to keep pace with improved technology, and make certain they were implemented. That was a start.

The most vital step we can take to reduce risk is to stop gain-of-function research. Just prohibit it. I was raised in the world of science. I have spent my entire career in laboratories. So I know scientists do not want the government regulating their work. They don't want barriers put in front of their research. Scientific independence is vital. The vast majority of scientists will fight against any perceived restrictions on their work. But right now, the actual dangers of gain-of-function research far outweigh the potential benefits. There are men and women actually trying to teach these viruses how to infect human tissue because they believe that will allow them to find a vaccine. That just is not smart.

The first gain-of-function experiments—although that was not the term used to describe them—took place almost a century ago when scientists working with bacteria began changing their genetic structure. Beginning in the 1960s, GoF techniques were used to further our understanding of DNA, our ability to combine DNA from different sources, and our ability to create recombinant DNA. Essentially, this work gave birth to the biotechnology industry. There was little danger in that work. But that changed at the beginning of the twenty-first century when virologists began manipulating influenza viruses, creating the present danger.

In addition to putting a moratorium on it in American laboratories, our country needs to stop funding countries, companies, and universities anywhere in the world doing gain-of-function research. If the German, or French, or British government any government—allows companies or universities to conduct this type of research, then boom! We cut off all funding from our government.

I am not naive. I understand that argument that stopping the good guys does not stop the bad guys. Of course that's true, but it certainly reduces the possibility of a pathogen escaping. There are only a limited number of scientists in the world capable of conducting this level of research. It isn't as simple as getting a gun. There is absolutely no reason to make it easier for anyone to do this work. An almost infinite number of books and movies have been based on the good guy stopping the mad scientist from creating this type of bio-weapon. But it no longer is just an exciting plot; it's a real possibility and we must do anything and everything possible to stop that from happening. If—if—if that research made us safer, and if it resulted in greater protection from our enemies, I would want to continue it with constraints. But there is no evidence it does that. And accidents do happen.

I also would love to reinstate the moratorium on publishing details of gain-of-function experiments. Among the most closely guarded secrets in American history was how to split the atom. It was given the highest secrecy classification. In 1953, Julius and Ethel Rosenberg were executed for giving that information to the Soviet Union. Gain-of-function technology has the potential of changing the world much more than an atom bomb. So why are the people performing these experiments allowed to share the details of their work? This is valuable information that could put our enemies on the path to creating new and deadly pathogens. I can accept publishing papers describing in general the experiments and the results, but not the details, not a how-to guide. Other researchers who need to know those details to further their own work can make contact. To me, this is simple logic.

Wherever the next pandemic comes from, we certainly aren't prepared to meet it. The Defense Department has bases around the world acting as our sentinels, our first line of defense, and they are able to put into instant action a national defense strategy. The military is supported by a robust network of private-sector contractors—Boeing, Northrup, the Rand Corporation, IBM, and others—working diligently to ensure the military has the weapons it needs to meet any traditional threat. Now we need to create the same layers of protection for biosecurity.

They don't exist. At least not on the level necessary to meet this threat. Our government should have an A-team responsible for our biosecurity. Right now that does not exist. We need an organized group of highly knowledgeable people with the power to make the decisions necessary to protect this country and enforce them. The Department of Defense already is stretched thin, so this should be grounded in the Department of Energy, which currently operates seventeen national labs. They have the expertise to oversee a massive program to build our capability to respond to biological threats. Just as defense contractors provide planes and ammunition, multiple contractors should be developing antiviral drugs, vaccines, diagnostic tests and equipment, medical devices, and cutting-edge medical supplies that might become necessary.

This is a lesson we should have learned. In the initial months of the Covid pandemic, we were woefully undersupplied. We needed everything from masks to ventilators. Vice President Pence had to ask Ford to take a car factory offline to produce ventilators. We can't be caught unprepared again.

It would be expensive to create that capability. It might cost $100 billion or more a year. But it is money invested in America's security and our individual safety. It will require a commitment from the government that presently does not exist—even after Covid. Normally the government funds some research but private companies profit from the sales of products. Preparedness is a hard sell. There isn't nearly as much profit in it. The defense department has to develop some sort of cost-benefit formula that serves both the public and private sectors.

We need it to engage the private sector because most of the science done in the public sector doesn't solve problems. It's information oriented rather than profit driven. Among the very first discussions we had in the Covid task force was how to produce and distribute a vaccine as quickly as possible. Some of our people advocated granting $15 billion to the NIH to pay for a crash project. They wanted to put everything else aside and spend whatever was necessary to create a vaccine. The guesstimate was that it could be done in three years—if we were lucky. We didn't have three years. Deborah Birx and I suggested that we should instead use that money to motivate the private sector. We got the Defense Department involved to distribute it, and we motivated the private sector to deliver a vaccine within a year.

Money motivates. As we pointed out, the NIH had not historically developed successful vaccines; that wasn't its expertise. But the private sector—the for-profit sector—had produced multiple lifesaving vaccines. That's certainly what happened to meet the AIDS epidemic. We were facing a new pathogen; it had evolved in nonhuman primates, and it had learned how to infect people. We started at the "what is this new thing?" stage. During those first few years, all my patients died pretty much within a year. Then private companies began developing new therapies, like AZT, that at least slowed down the progression of the disease. Within a decade, pharmaceutical companies had discovered medications that enabled people to live. And within another decade, we had simple nontoxic medicines that enabled infected people to live long lives and prevent the disease in others. We went from thirty pills a day that had only a small effect to a single pill that completely managed the virus. With the help of private companies, science did that.

We need to provide the financial support the private sector needs to invest in high-risk and expensive research to develop the countermeasures that are necessary. Our government could accomplish that by making it a priority to develop not one or two, but ten or twenty antivirals for avian flu, for Covid, for other viruses we know exist. We can prepare. We have the technology to create platforms that can be applied to meet emerging threats—just as we did with mRNA technology to create the Covid vaccines—so we don't have to live in helpless fear of the next infectious pathogen.

What we don't have is the commitment. And I predict we will regret that.

The most important tools we need to prevent the next pandemic are vaccines that don't yet exist. But there is significant and promising work being done. For years I doubted mRNA technology was going to work; I was wrong. And we're just learning how to use it to code for different amino acids. The long-term concept is that scientists will be able to create a specially designed molecule in the laboratory that can be injected into the body to make whatever medicines are needed. People wouldn't have to go to the drugstore to buy expensive medicines because their body would produce the medicine on demand. Like programming a computer, this technology can transform your body into a factory producing the necessary vaccine. AI may even speed up the process. It also might sequence the virus and tell us what molecules we want to make in response.

It is possible; it is a goal rather than a dream. Prototypes are already being used to fight certain types of cancer. The mRNA technology as it already exists will allow us to make a vaccine within weeks or a few months, That said, creating a vaccine is quite different from producing and distributing hundreds of millions of doses around the world

As vital as they will be, vaccines alone will not be enough. We know that flu vaccines have only been moderately successful, in large part because there are so many variants. In a good year, the flu vaccine is only 50 to 60 percent effective. During the COVID-19 pandemic, we had one of the most successful vaccine development programs in history; yet 1.2 million Americans died. That's not the definition of a success. That is evidence that vaccines aren't the answer to controlling respiratory pathogens. They need to be complemented by highly effective antivirals.

Antiviral drugs are the key to our security. We don't have highly effective antivirals.

The most promising treatments for the variety of bird flu vriuses are antiviral drugs. It's necessary for me to report that in recent years I have been working privately to develop a viable treatment for bird flu. We are hoping to follow the FDA's animal rule: If you can demonstrate a drug works safely in animals with the disease, you can move forward in human volunteers. We started by giving mice bird flu. The mice that weren't treated died; the mice given the drug lived. Then we repeated that with ferrets and got similar results. Then we gave a nonlethal dose of the virus to monkeys; those animals not given the drug lost weight and had significant viral loads in their lungs, but the drug prevented that from happening in our test animals. So we are on the right path. I suspect a number of other programs around the world are on that same path.

Right now, we have several drugs available that have shown some efficacy against viral diseases, even though they haven't been approved for that purpose. Personally, I'm a strong advocate for allowing physicians to use whatever works. I have absolutely no problem with physicians prescribing drugs off-label. At various times, I have done it myself, and I have seen some success.

I also believe we need to discuss the positives and negatives of off-label prescription. During Covid, for example, there was a limited amount of anecdotal evidence suggesting hydroxychloroquine was having some impact in fighting the disease. I had no idea if that was true or not. But doctors were using this drug, so I wanted them to be informed. I asked the editors of the *Morbidity and Mortality*

Weekly Report, MMWR, our weekly newsletter, to do a piece summarizing everything we knew about it. How was it being used? At what doses? What were the side effects? What had the response been? It was all very useful factual material. Some of it was positive, much of it was negative, but it gave physicians the information they needed to make rational decisions.

Immediately I got pummeled by my own people, who warned that I was advocating for the use of an unproven and perhaps ineffective drug for this application. That was exactly what I wasn't doing. My objective was to make accurate, up-to-date information available to prescribers. The debate was ongoing, and I wanted doctors to be sufficiently informed to make their own decisions.

That response was disappointing. Disheartening. Science is built on discussion and debate, not emotion and certainly not politics. Too often this debate broke down along political lines. The scientific data got lost in the turmoil. As Aeschylus wrote, "in war, truth is the first casualty."

I did not do the same thing for ivermectin because it had been in clinical use for a long time and most doctors had experience with it.

None of that changes the larger point: In war, you adapt the weapons you have to meet your needs. Doctors often know from their own clinical experience what therapies are effective in treating conditions. Other than protecting patients from crackpot remedies and scams, the government has no business in the doctor/patient relationship. My job was to make sure the medical community had all the factual information available to enable physicians to make informed decisions about what was best for their patients. My own clinical experiences taught me decades earlier that no bureaucrat in Washington knew what was better for my patients than I did. I had no interest or intention to interfere with that relationship.

While we have been diligently pursuing methods to prevent and treat outbreaks, the fact remains that we were caught essentially unprepared for COVID-19. We made mistakes. Damage was done. I was on the front lines of that war, and I learned some vitally important lessons. In addition to the scientific research currently being conducted, there are several other things we should do to correct those mistakes and to be better prepared for the next pandemic.

We need to begin by restructuring our public health system. Right now, there are numerous public health agencies. Thousands of them exist at the local, city, state, and national levels. Every jurisdiction has a public health department. That might be beneficial when dealing with local matters, but not when the entire nation is facing a threat. In that situation, we have to speak with one voice, we have to communicate one message. The way the system exists now, Americans are often getting confusing or even conflicting messages. No one knows who is in charge because many different people have at least partial responsibility.

At least part of the reason for this problem is that different departments and agencies in the federal government have overlapping responsibilities. That's just the way the system has evolved. Various elements of the same issue might be the responsibility of the NIH, the CDC, or the FDA. It could also be a national security question: Was Covid bioterrorism? If the answer is yes, even more agencies would get involved. A simple example: Whose job was it to ensure healthcare workers had masks to protect them? The federal government? state health officials? Hospital administrators? The right answer is all of them. There should be a pipeline from the government directly into the clinic with clearly delineated responsibilities to make certain we have the materials and equipment needed to meet the challenge.

That didn't exist. The result was we did not have the masks we needed, and nobody really knew how to solve that problem. The military works efficiently because it maintains a strict chain of command. That ensures that leadership decisions are effectively communicated to troops, that they are carried out, and that people on all levels have responsibility for their actions. That doesn't exist in our public health system. We are caught in a morass of many agencies, sometimes with competing agendas, that all have decision-making authority.

One result of that, as we witnessed during Covid, is that Americans have lost trust in the public health sector. They believe the government has a hidden or disguised agenda. That mixed messaging was confounded even more by the abundance of internet experts. And people died because of it.

Without question, we need to restructure the nation's public health agencies into an orderly top-down system in which policy-making responsibilities are clearly delineated and accountability is visible. I think we need to have one individual who is considered credible, rather than multiple people who may be putting their own spin on the story.

But here's the rub: Even if there had been a single agency or person speaking for the administration and the collective public health agencies, too many people wouldn't have paid attention to that person.

People no longer trust public health officials. We got the initial message wrong. COVID-19 was not SARS, it was not MERS. It was unique, different, new. We did not understand that asymptomatic people could be infectious. Then the light bulb went on: In March 2020, we learned that asymptomatic transmission was a dominant feature and that the approach we had been taking wasn't going to work. This wasn't anyone's fault. No one lied. That's just the way science works. We had taken the immediate actions necessary to save lives based on what we knew or believed at the time. Then we learned we were wrong. We shifted gears to respond to this new information. We didn't hide it; we didn't broadcast it either. But Americans saw for themselves that what we had told them wasn't true. From the beginning, the government's credibility was damaged.

That was followed by a stream of inaccurate, incomplete, or misleading information. There was no scientific data behind the social distancing recommendation of six feet. Being infected with Covid did not prevent you from being infected again and again. There was no real possibility of herd immunity. The promise, fully vaccinated, was a myth; the vaccines did not prevent infection. Vaccines can have side effects. People of all ages and physical conditions were not equally endangered by the disease. They also weren't equally healthy enough to be taking the vaccine. Older people, obese people, people with diabetes or heart conditions fared differently than younger, healthier individuals. All of these problems were magnified by the growing political divisions in this country, as competing politicians cherry-picked facts for their own benefit.

Many of the ostensibly life-changing actions taken in support of dubious science made the situation even worse. From keeping schools and businesses closed longer than necessary, to unnecessarily mandating vaccines, to firing first responders who refused to be vaccinated, the response created serious problems that could have been avoided.

The effect of all of it—all of the noise, all of the accusations, all of the anger and bitterness—was this loss of trust in public health agencies. Many Americans no longer believe these agencies are most concerned about protecting their health. Many Americans no longer believe public health officials like me. If I say something that doesn't support their preexisting opinion—regardless of the science—they attack me. For the entirety of my career, I have not been a partisan. I have been a messenger for science. Science, unlike politics, doesn't lie. If I say that science has unequivocally determined that vaccines do not cause autism, people will not only disagree with me, the messenger, they will attack me verbally and in print. Often, those attacks can get quite nasty. This stifling environment is not hospitable to the research and the difficult conversations that we need to have to prevent the next pandemic.

Evidence of the damage that this loss of confidence can cause was the 2024–2025 measles outbreak in the Southwest. Science says the measles vaccine works. Some Americans don't trust it nor the authorities who advocate it. So they refuse to vaccinate their children. Measles spreads. It's a devastating equation.

The ability to communicate with the American public during a crisis is critical. And that was a casualty of COVID-19. Somehow that trust has to be restored before the next pandemic begins. If I had a magic wand, I would wave it and that trust would be restored. I don't, and I'm not certain how it can be done. It's a very high hurdle. But I know where to begin: Tell the truth, the whole truth, and nothing but the truth.

Just tell the truth—even when the truth is, "we don't know." In those early days, I often admitted that we didn't have an answer yet. I was criticized for that too. I was told, you have to have an answer. I agreed. My answer was, I don't know the answer.

You don't build trust and confidence by providing inaccurate or incomplete information. Our healthcare leaders need to be candid with America. "I don't know yet" or "This is what I think based on the information we have" are both perfectly fine answers. And when we're wrong, we have to admit it. We can explain why we were wrong, we can show the evolution of the science, and we can present the new facts as they emerge. Ironically, being able to admit it when you are wrong is actually a way of building future trust.

As has become obvious, my opinions, my beliefs, and my convictions are shaped and reshaped by the data. When dealing with the public, the data on which a statement is based has to be made available to the public. When we advise people that staying six feet away from others will prevent infection, they need to know how we determined that figure. Why six feet? Is that based on a reliable study? And if the data don't exist, maybe we shouldn't offer the recommendation.

The days of believing authorities because they are in an authoritative position are over. The way to convince people to follow recommendations is to provide evidence that the recommendations are not arbitrary or politically motivated, but based on science. We need to provide context. We need to admit when we are wrong. We need to be willing to change as the data changes. Policy recommendations are never going to be a perfect science. But, in our commitment to science and telling the American public the truth, we will save lives.